A COURSE IN QUANTUM MECHANICS

Nandita Rudra

Former Faculty, Department of Physics, University of Kalyani,
Kalyani, WB, India

CRC Press
Taylor & Francis Group
Boca Raton London New York

CRC Press is an imprint of the
Taylor & Francis Group, an **informa** business

Levant Books
India

CRC Press
Taylor & Francis Group
6000 Broken Sound Parkway NW, Suite 300
Boca Raton, FL 33487-2742

First issued in paperback 2023

© 2020 by Nandita Rudra and Levant Books

CRC Press is an imprint of the Taylor & Francis Group, an informa business

No claim to original U.S. Government works

ISBN 13: 978-1-03-265410-2 (pbk)
ISBN 13: 978-0-367-34429-0 (hbk)
ISBN 13: 978-0-429-32576-2 (ebk)

DOI: 10.1201/9780429325762

This book contains information obtained from authentic and highly regarded sources. Reasonable efforts have been made to publish reliable data and information, but the author and publisher cannot assume responsibility for the validity of all materials or the consequences of their use. The authors and publishers have attempted to trace the copyright holders of all material reproduced in this publication and apologize to copyright holders if permission to publish in this form has not been obtained. If any copyright material has not been acknowledged please write and let us know so we may rectify in any future reprint.

Except as permitted under U.S. Copyright Law, no part of this book may be reprinted, reproduced, transmitted, or utilized in any form by any electronic, mechanical, or other means, now known or hereafter invented, including photocopying, microfilming, and recording, or in any information storage or retrieval system, without written permission from the publishers.

Trademark notice: Product or corporate names may be trademarks or registered trademarks, and are used only for identification and explanation without intent to infringe.

Publisher's Note
The publisher has gone to great lengths to ensure the quality of this reprint but points out that some imperfections in the original copies may be apparent.

Print edition not for sale in South Asia (India, Sri Lanka, Nepal, Bangladesh, Pakistan or Bhutan)

Library of Congress Cataloging-in-Publication Data
A catalog record has been requested

Visit the Taylor & Francis Web site at
http://www.taylorandfrancis.com

and the CRC Press Web site at
http://www.crcpress.com

In loving memory of
Prasantakumar Rudra
October 7, 1939—May 24, 2017

Preface

This book is the outcome of the two semester course of lectures that we have been delivering for a number of years to the first year M.Sc. students at the University of Kalyani, in West Bengal. The material contained in it is appropriate for M.Sc and advanced under graduate (senior level) students covering the syllabi of most of the universities of India and abroad.

Instead of following the historical development of quantum mechanics the theory has been built up starting from a number of basic postulates as described in chapter 1.

Representation theory comes next in chapter 2. It includes changes of representation and the concept of unitary transformation which is very basic in the formulation of the theory. After considering spatial translation and momentum we go over to the time evolution of quantum systems which finally gives rise to the Schrodinger equation of motion. Feynman's Path Integral formalism then follows in chapter 5. Chapter 6 contains the application of quantum mechanics to one dimensional problems. More realistic three dimensional cases are dealt with in chapter 10.

Rotation and angular momentum are discussed in great detail in chapters 7, 8, and 9.

Symmetry plays and important role in quantum mechanics, perhaps more so than in classical mechanics. This is covered in chapter 11.

Since most of the physical problems cannot be exactly solved, different approximate methods are developed which constitute chapter 12.

Time dependent approximate methods and scattering processes are discussed elaborately in chapters 13, 14 and 15.

Relativistic quantum mechanics is introduced in the last chapter ending at the doorstep of quantum field theory.

I am undebted to Dr. G Speisman of the Florida State University, USA, whose lectures I attended as a graduate student. These lecture notes have been of help to me.

I would like to thank my former colleagues Professor Ratanlal Sarkar for many helpful discussions and Professor Siddhartha Ray for supplying me with books and other materials. Professor Ray also managed to find time from his busy schedule to oversee the entire process of publication of this book. I am extremely grateful to him.

This book was written when we were visiting our sons Angsuman and Archisman in America. They took care of us and provided a perfect ambience for me to finish the book. I am so proud of them.

Finally my husband Professor Prasantakumar Rudra motivated me to write this book and painstakingly prepared the La Tex version of the manuscript in camera ready form. This book would not have been possible without him.

Kolkata, India   Nandita Rudra
2018

Contents

Notations & Fundamental Constants

α^{-1}	Inverse Fine Structure Constant = 137.035 999 679(50)
β	$\frac{1}{k_B T}$
$\delta(x)$	Dirac delta function of x
μ_B	Bohr Magneton, $\frac{e\hbar}{2m_e}$ = 9.274 009 15(23) $\times 10^{-24}$ JT^{-1}
$\boldsymbol{\sigma}$	Pauli Matrix
$\hat{\mathbf{a}}$	Unit vector along $\mathbf{a}$
$\left[\hat{A}, \hat{B}\right]$	Commutator of two quantum mechanical operators $\hat{A}$ & $\hat{B}$
$[A, B]_{\text{PB}}$	Poisson bracket of two classical functions A & B
$\mathbf{B}$	Magnetic Field
c	Speed of light in vacuum = 2.997 924 58 $\times 10^8$ ms^{-1}
e	Elementary Charge = 1.602 176 487(40) $\times 10^{-19}$ C
h	Planck's Constant = 6.626 068 96(33) $\times 10^{-34}$ Js
$\hbar$	$\frac{h}{2\pi}$ = 1.054 571 628(53) $\times 10^{-34}$ Js
$\hat{H}$	Hamiltonian operator
$\Im(z)$	Imaginary Part of complex number z
k_B	Boltzmann Constant = 1.380 650 4(24) $\times 10^{-23}$ JK^{-1}
m	Mass of Electron = 9.109 382 15(45) $\times 10^{-31}$ kg
m_p	Mass of Proton = 1.672 621 637(83) $\times 10^{-27}$ kg
n	Number Density
$\hat{\mathcal{N}}$	Number operator
N_A	Avogadro Constant = 6.022 141 79(30) $\times 10^{23}$ mol^{-1}
$\Re(z)$	Real Part of complex number z
Tr	Trace Operator

Chapter 1

Basic Concepts & Formulation

1.1 Introduction

Newtonian Mechanics, theory of Elasticity and Fluid Dynamics, together with Maxwellian Electrodynamics are basically what constitute Classical Physics. Classical Physics was developed and successfully applied to describe the motions of macroscopic bodies like planetary systems and other every day phenomena. It also described electromagnetic field and its interaction with matter. The development of classical physics was almost complete by the early twentieth century. With the discovery of subatomic particles, radioactivity and X-rays and also the accumulation of data on the spectrum of Black Body Radiation and other spectroscopic studies, it became imperative to invoke new concepts and ideas. Thus Quantum Mechanics, the fundamental theory of the physical world was formulated.

Based on the early work by Planck, Einstein, Bohr, de Broglie, Sommerfeld and others, Quantum Mechanics was developed during 1925 - 1928 by Heisenberg, Born, Jordan, Schrödinger, Dirac, Pauli and others. All experiments performed during the last 80 years have indicated the correctness of Quantum Mechanics.

Quantum Mechanics has brought about profound change in our thinking about the description of natural phnomena, particularly those of the subatomic world. Though it is based on concepts radically different from those of classical physics, quantum mechanics yields the laws of classical physics in the macroscopic limit.

1.2 Measurements

In classical mechanics the dynamical state of a system is determined from the knowledge of the dynamical variables like the coordinates and momenta of the constituents at every moment of time. All such variables can, in principle, be simultaneously measured with infinite precision.

In Quantum Mechanics, on the other hand, measuremnt process itself, *i.e.* the presence of the measuring apparatus between the system and the observer modifies the state of the system in an unpredictable way such that the subsequent measurements may yield different values for the same variable. The result of the measurement in quantum mechanics is probabilistic rather than deterministic as in classical mechanics. This probabilistic nature of measurement is an inherent property of quantum systems unlike statistical probability which arises due to the lack of knowledge of all the initial values of the variables that define the state.

This means that quantum mechanics predicts the number of times n that a particular result will be obtained, when a large number N of measurements are carried out on a collection of *identical* and *independent* systems that are *identically* prepared. Such a collection is called an **Ensemble**.

1.3   Basic Postulates

We shall start by enunciating and explaining the *basic postulates and axioms* that are necessary for formulating the *new mechanics*. These postulates are self-evident truths that cannot be *proved* but have to be accepted for building up any new theory. If the theory thus built up is successful not only in explaining the known experimental results but also in predicting new ones, then these *postulates* become *laws*. Thus we have the *Newton's Laws* in classical mechanics, the *Maxwell's Equaions* in classical electrodynamics. In statistical physics we similarly have the postulates of *equal a priori probability* and *ergodic hypothesis*.

(i)   Postulate 1. State of the System

We shall use the *ket* and *bra* vector notations developed by Dirac. These elegant notations are compact, sufficiently general and extremely powerful as will become evident as we proceed.

The physical state of a system is denoted by a *ket vector* $|\alpha\rangle$ in a linear and complex vector space which is a *mathematical space* with dimension depending on the nature of the physical system under consideration. The vector space in infinite dimension is known as *Hilbert Space*.The state index α denotes the set of values or *quantum numbers* of the physical quantities which define the state of the system. The *ket* $|\alpha\rangle$ is postulated to contain *all* the informations that can be known about the system. Also

$$c|\alpha\rangle = |\alpha\rangle c, \ c = \text{arbitrary complex number,} \tag{1.1}$$

is postulated to represent the same state $|\alpha\rangle$. When c is zero the result is the *Null ket*.

(ii) Postulate 2. Superposition of States.

This postulate is one of the corner stones of quantum mechanics. According to this postulate if there are more than one states of the system then the linear combination will also represent another state of the system. For instance if $|\alpha\rangle$ and $|\beta\rangle$ are two possible states, then

$$c_\alpha|\alpha\rangle \ + \ c_\beta|\beta\rangle \ \equiv \ |\gamma\rangle, \text{ where } c_\alpha \text{ and } c_\beta \text{ are arbitrary complex numbers,} \tag{1.2}$$

will represent another state of the system. Thus one can add up states to generate new states by superposition. The physical vectors in *Euclidian Space* have the property that they can be added up to form new vectors. Analogously we have designated the states $|\alpha\rangle$ as *ket vectors* in a complex and linear mathematical space. Since from Eq. (1.1) $c|\alpha\rangle$ and $|\alpha\rangle$ are the same state when $c \neq 0$, only the 'direction' and not the magnitude of the ket vector is of significance.

BRA VECTORS, BRA SPACE, INNER PRODUCT

We now introduce the *bra vectors* and the *bra space*. According to Dirac every ket vector $|\alpha\rangle$ is associated with a *bra vector*, denoted by $\langle\alpha|$, by some *conjugation process*. The association between $|\alpha\rangle$ and $\langle\alpha|$, we call *dual correspondence*. The *Bra Vectors* are introduced such that the complete *bracket* notation $\langle\beta|\alpha\rangle$ which is defined as the *inner product* of a *ket* $|\alpha\rangle$ and a *bra* $\langle\beta|$ will be in general a complex number. The inner product is postulated in analogy to the *scalar product* $\mathbf{a} \cdot \mathbf{b}$, which is a scalar quantity, of two physical vectors $\mathbf{a}$ and $\mathbf{b}$. According to Dirac a bra vector is completely defined when its inner product with *all* the kets in the ket space is given. If this inner product of $\langle\beta|$ is *zero* for *all* the kets $|\alpha\rangle$, then $\langle\beta|$ is a *Null* bra vector.

$$\text{If } \langle\beta|\alpha\rangle \ = \ 0, \ \text{ for all } |\alpha\rangle, \tag{1.3}$$
$$\text{then } \langle\beta| \ = \ 0, \ \text{ Null Bra Vector.} \tag{1.4}$$

Thus with every ket space there is an associated bra space spanned by bra vectors. We write

$$|\alpha\rangle \ \overset{\text{Dual Correspondence}}{\longleftarrow \cdots \cdots \longrightarrow} \ \langle\alpha|,$$
$$|\alpha\rangle \ \overset{\text{DC}}{\longleftrightarrow} \ \langle\alpha|, \tag{1.5}$$

which we may call *Hermitian Conjugation* and write

$$\langle\alpha| \;=\; [|\alpha\rangle]^{\dagger} \text{ and } \langle\alpha|^{\dagger} \;=\; |\alpha\rangle.$$

in general

$$|\alpha\rangle + |\beta\rangle \;\overset{\text{DC}}{\longleftrightarrow}\; \langle\alpha| + \langle\beta|, \tag{1.6}$$

$$c_\alpha|\alpha\rangle + c_\beta|\beta\rangle \;\overset{\text{DC}}{\longleftrightarrow}\; c_\alpha^*\langle\alpha| + c_\beta^*\langle\beta|, \tag{1.7}$$

where c_α and c_β are complex numbers.

Having defined the inner product

$$(\langle\beta|) \cdot (|\alpha\rangle) \;=\; \langle\beta|\alpha\rangle, \tag{1.8}$$

we further **postulate** that

$$\langle\beta|\alpha\rangle \;=\; \langle\alpha|\beta\rangle^* \tag{1.9}$$

Thus the numbers $\langle\beta|\alpha\rangle$ and $\langle\alpha|\beta\rangle$ are complex conjugate of each other. Though the *inner product* is analogous to *scalar product* of two vectors, but unlike it the inner product is *non-commutative, i.e.*

$$\langle\beta|\alpha\rangle \;\neq\; \langle\alpha|\beta\rangle. \tag{1.10}$$

Just as $\mathbf{a} \cdot \mathbf{b} = a_x b_x + a_y b_y + a_z b_z = ab\,\cos\theta$ is the overlap of the vector $\mathbf{a}$ with vector $\mathbf{b}$, $\langle\beta|\alpha\rangle$ represents the overlap integral of $\langle\beta|$ and $|\alpha\rangle$; which will become clear once the representation of $|\beta\rangle$ and $|\alpha\rangle$ in terms of the complete set of basis is introduced.

From Eq. (1.9) follows

$$\langle\alpha|\alpha\rangle \;=\; \langle\alpha|\alpha\rangle^* \;=\; \text{Real}. \tag{1.11}$$

We further **postulate** that

$$\langle\alpha|\alpha\rangle \;\geq\; 0, \tag{1.12}$$

where the equality holds if and only if $|\alpha\rangle$ is the *Null ket*. We shall call $\sqrt{\langle\alpha|\alpha\rangle}$ the *Norm* of the ket $|\alpha\rangle$ and the postulate of positive definiteness of norm is essential for the probabilistic interpretation of quantum mechanics. Two kets $|\alpha\rangle$ $|\beta\rangle$ are said to be *orthogonal* if

$$\langle\alpha|\beta\rangle \;=\; 0. \tag{1.13}$$

The *normalized* ket $|\tilde{\alpha}\rangle$ is given by

$$|\tilde{\alpha}\rangle \;=\; \frac{|\alpha\rangle}{\sqrt{\langle\alpha|\alpha\rangle}}. \tag{1.14}$$

The norm $\sqrt{\langle\alpha|\alpha\rangle}$ is analogous to the magnitude $[\mathbf{a}\cdot\mathbf{a}]^{\frac{1}{2}} = |\mathbf{a}|$ of the vector $\mathbf{a}$. Since $|\alpha\rangle$ and $c|\alpha\rangle$ represent the same physical state, we shall henceforth use the normalized kets Eq. (1.14) to describe a state. The normalized ket vectors can yet be multiplied by a phase factor $e^{i\gamma}$, with real γ with *modulus* unity, before it is completely specified.

(iii)   Postulate 3. Operators for Dynamical Variables

Every *Dynamical Variable* A will be represented by a linear operator $\hat{A}$ which operates on a ket $|\alpha\rangle$ to transform it to another ket $|\gamma\rangle$

$$\hat{A}(|\alpha\rangle) = \hat{A}|\alpha\rangle = |\gamma\rangle. \tag{1.15}$$

Then

$$\langle\gamma| = (|\gamma\rangle)^{\dagger} \equiv \langle\alpha|\hat{A}^{\dagger}, \tag{1.16}$$

where $\hat{A}^{\dagger}$ defined to be *Hermitian Adjoint* of $\hat{A}$ which acts on the bra $\langle\alpha|$ from the right. Since $\hat{A}$ has to be *linear* we have

$$\hat{A}[c_{\alpha}|\alpha\rangle + c_{\beta}|\beta\rangle] = c_{\alpha}\hat{A}|\alpha\rangle + c_{\beta}\hat{A}|\beta\rangle. \tag{1.17}$$

OUTER PRODUCT OF STATES.

We define the *Outer Product* of $|\beta\rangle$ and $|\alpha\rangle$

$$(|\beta\rangle)\cdot(\langle\alpha|) \equiv |\beta\rangle\langle\alpha|. \tag{1.18}$$

Unlike the inner product $\langle\beta|\alpha\rangle$ which is a number, it can be shown that the outer ptoduct Eq. (1.18) is an operator. To show this we use after Dirac the *associative axiom* of multiplication. Just as the multiplications between operators are associative, Dirac postulated this property holds good for any legal multiplication (*i.e.* multiplications that are allowed) among kets, bras and operators. Thus

$$(|\beta\rangle\langle\alpha|)\,|\gamma\rangle = |\beta\rangle\cdot\langle\alpha|\gamma\rangle = c_{\alpha\gamma}|\beta\rangle, \tag{1.19}$$

where $c_{\alpha\gamma} = \langle\alpha|\gamma\rangle = $ a number.

It is as if $|\beta\rangle\langle\alpha|$ rotated $|\gamma\rangle$ in the direction of $|\beta\rangle$. Similarly

$$\langle\gamma|\cdot(|\beta\rangle\langle\alpha|) = \langle\gamma|\beta\rangle\cdot\langle\alpha| = \langle\alpha|c_{\gamma\beta} \tag{1.20}$$

HERMITIAN ADJOINT OF $|\beta\rangle\langle\alpha|$.

We have defined Hermitian Adjoint operator in Eq. (1.16). Let

$$\hat{X} \equiv |\beta\rangle\langle\alpha|, \tag{1.21}$$

$$\hat{X}|\gamma\rangle = (|\beta\rangle\langle\alpha|) \cdot |\gamma\rangle = |\beta\rangle \cdot \langle\alpha|\gamma\rangle$$
$$= c_{\alpha\gamma}|\beta\rangle . \tag{1.22}$$

Then

$$\langle\gamma|\hat{X}^\dagger = \langle\alpha|\gamma\rangle^* \cdot \langle\beta| = \langle\gamma|\alpha\rangle\langle\beta|$$
$$= \langle\gamma| \cdot (|\alpha\rangle\langle\beta|) . \tag{1.23}$$

Thus

$$\hat{X}^\dagger = |\alpha\rangle\langle\beta|, \text{ where } \hat{X} = |\beta\rangle\langle\alpha|. \tag{1.24}$$

$\hat{X}^\dagger$ is called the *Hermitian Conjugate* of $\hat{X}$.
An operator is said to be *Hermitian* if it is equal to its Hermitian Conjugate.

i.e. if $\hat{X} = \hat{X}^\dagger$, then $\hat{X}$ *is Hermitian*.

Since

$$\langle\beta|\hat{X}|\alpha\rangle = \langle\beta| \cdot \left(\hat{X}|\alpha\rangle\right) = \left[\left(\langle\alpha|\hat{X}^\dagger\right) \cdot |\beta\rangle\right]^*, \tag{1.25}$$

if $\hat{X} = \hat{X}^\dagger$,   then,

$$\langle\beta|\hat{X}|\alpha\rangle = \langle\alpha|\hat{X}|\beta\rangle^* \tag{1.26}$$

and,   in particular

$$\langle\beta|\hat{X}|\beta\rangle = \langle\beta|\hat{X}|\beta\rangle^* = \text{Real}$$

In other words the diagonal matrix elements of Hermitian Operators are real.

(iv)   Postulate 4. Eigenvalues and Eigenvectors of Operators

When an operator acts on a ket $|\alpha\rangle$ it does not, in general, give rise to the same ket $|\alpha\rangle$ times a constant. However, for an operator $\hat{A}$, there *may* exist a particular class of kets known as *eigenkets* of $\hat{A}$ satisfying the following equation

$$\hat{A}|a_n\rangle = a_n|a_n\rangle, \quad n = 1, 2, \cdots , \tag{1.27}$$

where $\{a_1, a_2, \cdots \}$ are just numbers, called the *eigenvalues* and Eq. (1.27) is known as *Eigenvalue Equation*. The kets $|a_1\rangle$, $|a_2\rangle$, $\cdots$, are known as *Eigenkets*. The totality of the numbers $\{a_1, a_2, \cdots ,\}$ is called the *Spectrum of Eigenvalues*.

According to Dirac a measurement always causes the physical system to be thrown into one of the eigen states of the dynamical variable being measured and will yield the corrensponding eigenvalue. It is postulated that the only result of a precise measurement of the dynamical variable A is one of the eigenvalues a_n of the linear operator $\hat{A}$, associated with A. This postulate has to do with measurement on a quantum system.

(v) Postulate 5. Hermitian Operator.

According to the previous postulate measurement yields eigenvalues of an operator corresponding to a physical *observable*. Since physical obervables are neccessarily real quantities we need to represent them by such operators that have real eigenvalues.

We shall now prove that Hermitian operators have real eigenvalues and the eigenkets are orthonormal for non-degenerate solutions.

The eigenvalue equation for a Hermitian operator $\hat{A}$ is

$$\hat{A}\,|a_m\rangle \;=\; a_m\,|a_m\rangle \;\; \text{so that} \;\; \langle a_n|\hat{A}|a_m\rangle \;=\; a_m\langle a_n|a_m\rangle \tag{1.28}$$

$$\langle a_n|\hat{A}^\dagger \;=\; a_n^*\langle a_n| \;=\; \langle a_n|\hat{A} \;\; \text{so that} \;\; \langle a_n|\hat{A}|a_m\rangle \;=\; a_n^*\langle a_n|a_m\rangle, \tag{1.29}$$

and thus

$$\left(a_m \;-\; a_n^*\right)\langle a_m|a_n\rangle \;=\; 0. \tag{1.30}$$

Thus

$$\text{for } n = m, \; a_n = a_n^* = \text{ real} \quad \text{and} \quad \text{for } n \neq m, \; \langle a_n|a_m\rangle \;=\; 0. \tag{1.31}$$

The equation in Eq. (1.31) corresponds to non-degerate case. If the solutions are degenerate, *i.e.* when there are more than one distinct eigenkets for a particular eigenvalue, then orthogonality can be achieved by taking a suitable linear combination of the degenerate kets, that we leave as an exercise. So in all cases we can write Eq. (1.31) as follows

$$\langle a_n|a_m\rangle \;=\; \delta_{n.m}. \tag{1.32}$$

This is the *orthonormality condition*.

Thus from Eq. (1.31) we find that the eigenvalues of Hermitian operators are real. A sufficient but not necessary condition for an operator to have real eigenvalues is for it to be Hermitian. We *postulate* that the linear operators associated with observables are Hermitian.

(vi) Postulate 6. Basis Vectors and Completeness Condition.

We now *postulate* that the the eigenkets $\{|a_1\rangle,\ |a_2\rangle,\ \cdots\ \}$ of the Hermitian operator $\hat{A}$ corresponding to the observable A form a *complete set* spanning the whole of the ket space, such that an arbitrary ket $|\alpha\rangle$ of the system can be uniquely expanded in terms of this complete set. In other words the eigenkets $\{|a_1\rangle,\ |a_2\rangle.\ \cdots\ \}$ form a basis of the ket space. Then we can write for any arbitrary ket $|\alpha\rangle$

$$|\alpha\rangle\ =\ \sum_n c_{\alpha n}|a_n\rangle,\ \text{where}\ c_{\alpha n}\ =\ \langle a_n|\alpha\rangle \tag{1.33}$$

are the expansion parameters and are in general complex numbers.

Also since

$$|\alpha\rangle\ =\ \sum_n |a_n\rangle\langle a_n|\alpha\rangle \tag{1.34}$$

is true for any arbitrary ket $|\alpha\rangle$, we have

$$\sum_n |a_n\rangle\langle a_n|\ =\ \hat{I} \tag{1.35}$$

an *unit operator*. Eq. (1.35) gives the *completeness criterion* for the set $\{|a_1\rangle,\ |a_2\rangle,\ \cdots\ \}$ and is also known as the *closure property*. [1]

PROJECTION OPERATOR

Comparing Eq. (1.33) and Eq. (1.35) we see that the operator $|a_n\rangle\langle a_n|$ projects $|\alpha\rangle$ along $|a_n\rangle$ and is thus called the *projection operator*, defined as:

$$\hat{\Lambda}_n\ =\ |a_n\rangle\langle a_n|, \tag{1.36}$$
$$\hat{\Lambda}_n^2\ =\ \hat{\Lambda}_n,$$
$$\sum_n \hat{\Lambda}_n\ =\ \sum_n |a_n\rangle\langle a_n|\ =\ \hat{I} \tag{1.37}$$
$$=\ \text{Totality of all projections}$$

The unit operator so defined is extremely important as it can be inserted anywhere in a chain of multiplication of kets, bras and operators to derive useful results. For example

$$\langle\alpha|\alpha\rangle\ =\ \langle\alpha|\left(\sum_n |a_n\rangle\langle a_n|\right)|\alpha\rangle \tag{1.38}$$
$$=\ \sum_n \langle\alpha|a_n\rangle\langle a_n|\alpha\rangle\ =\ \sum_n c_{\alpha n}c_{\alpha n}^*\ =\ \sum_n |c_{\alpha n}|^2$$
$$=\ 1, \tag{1.39}$$

for a normalized ket $|\alpha\rangle$.

---

[1]See Appendix A.1 for discussion on completeness condition.

MEASUREMENT OF OBSERVABLE A

We now make a useful observation relating to the measurement of the observable A. If the state vector of the system is one of the eigenkets, *i.e.* if the system is in an eigenstate $|a_n\rangle$ of $\hat{A}$, then the measurement of A will yield the eigenvalue a_n with definiteness. On the other hand if the state vector is an arbitrary ket $|\alpha\rangle$, then the measurement of $\hat{A}$ will yield any one of the eigenvalues a_1, a_2, $\cdots$. The probability of obtaining a particular value a_n is postulated to be given by $|\langle a_n|\alpha\rangle|^2$ or $|c_{\alpha n}|^2$ where $c_{\alpha n}$ is the expansion parameter in Eq. (1.33). The parameter $c_{\alpha n}$ is known as the *probabilty amplitude* for the state $|a_n\rangle$ in $|\alpha\rangle$.

WHAT IS AN OBSERVABLE ?

According to Dirac not all dynamical variables are observables. If any variable is such that the corresponding operator does not possess a complete set of eigenkets and a system ket is not expressible in terms of them then such a variable cannot be called an observable in quantum mechanics.

(vii)   Expectation Value

This is defined as

$$\langle\hat{A}\rangle \;\equiv\; \text{Expectation value of } \hat{A} \;=\; \langle\alpha|\hat{A}|\alpha\rangle. \tag{1.40}$$

for a normalized state $|\alpha\rangle$. It can be shown that repeated measurements of an observable A or measurement on an ensemble of identical systems will yield the expectation value.

$$
\begin{aligned}
\langle\alpha|\hat{A}|\alpha\rangle &= \langle\alpha|\left(\sum_n |a_n\rangle\langle a_n|\right)\cdot\hat{A}\cdot\left(\sum_m |a_m\rangle\langle a_m|\right)|\alpha\rangle \tag{1.41}\\
&= \sum_{n.m}\langle\alpha|a_n\rangle\langle a_n|\hat{A}|a_m\rangle\langle a_m|\alpha\rangle \\
&= \sum_{n,m}\langle\alpha|a_n\rangle a_m\langle a_n|a_m\rangle\langle a_m|\alpha\rangle \\
&= \sum_{n,m}\langle\alpha|a_n\rangle a_m\,\delta_{n,m}\langle a_m|\alpha\rangle \\
&= \sum_n\langle\alpha|a_n\rangle a_n\langle a_n|\alpha\rangle \;=\; \sum_n a_n|\langle a_n|\alpha\rangle|^2, \tag{1.42}
\end{aligned}
$$

which is the weighted average of the eigenvalues a_n, hence this value will be repro-
duced on repeated measurement of A when the system is in a state $|\alpha\rangle$.

Problem 1.1 An observable A has only *two normalized eigenstates* $|\alpha_1\rangle$ and $|\alpha_2\rangle$
such that

$$A|\alpha_1\rangle = \alpha_1|\alpha_1\rangle, \text{ and } A|\alpha_2\rangle = \alpha_2|\alpha_2\rangle.$$

(i). Prove that $|\alpha_1\rangle$ and $|\alpha_2\rangle$ are orthogonal to each other.

(ii). What is their *completeness condition?*

(iii). Obtain the expecttion value of A in the state

$$|\phi\rangle = \sqrt{\frac{1}{3}}|\alpha_1\rangle + \sqrt{\frac{2}{3}}|\alpha_2\rangle.$$

Problem 1.2 Condider the operator

$$\hat{\Omega} = |\phi_1\rangle\langle\phi_2| + |\phi_2\rangle\langle\phi_1|$$

where $|\phi_1\rangle$ and $|\phi_2\rangle$ are orthonormal.

(i). Is $\hat{\Omega}$ a projection operator?

(ii). What is the completeness condition of $|\phi_1\rangle$ and $|\phi_2\rangle$?

Problem 1.3 For any state $|\psi\rangle$ obtain the eigenvalue and eigenvector of the oper-
ator $|\psi\rangle\langle\psi|$.

Problem 1.4 The state of a system is given by

$$|\alpha\rangle = C_1|\epsilon_1\rangle + C_2|\epsilon_2\rangle,$$

where $|\epsilon_1\rangle$ and $|\epsilon_2\rangle$ are the energy eigenstates with energy eigenvalues ϵ_1 and ϵ_2,
C_1 and C_2 being constants.

(i). What is the probability of getting an energy value ϵ_1 on a large number of
independent energy measurements in the state $|\alpha\rangle$?

(ii). Obtain the expectation value of the Hamiltonian $\langle\alpha|H|\alpha\rangle$.

Chapter 2

Representation Theory

2.1 Elements of Representation Theory

In the first course of quantum mechanics one defines a wavefunction $\psi_\alpha(\xi, t)$ to describe the state of the system. $\psi_\alpha(\xi.t)$ is a function of the coordinates ξ at a particular instant of time t. The parameters α denote the quantum numbers of the physical quantities having well-defined values which describe the state ψ.

In Dirac notations we denote the state of the system by a ket vector $|\alpha\rangle$ defined in a mathematical vector space which is linear and complex. This space is so constructed that the eigenkets $\{|a_1\rangle, |a_2\rangle, \cdots \}$ of an Hermitian operator $\hat{A}$ form a complete and orthonormal set spanning the whole of this space called the *Hilbert Space* in general. This complete set then forms the basis vectors of the space and an arbitrary physical state $|\alpha\rangle$ can be expressed, as in Eq. (1.33), in the basis $\{|a_n\rangle\}$:

$$|\alpha\rangle \;=\; \sum_n |a_n\rangle\langle a_n|\alpha\rangle \;=\; \sum_n c_{\alpha n} |a_n\rangle,$$

with $c_{\alpha n} = \langle a_n|\alpha\rangle$ forming the 'coordinates' of the state $|\alpha\rangle$ in the basis $|a_n\rangle$. This set of numbers is called the *wavefunction in $\hat{A}$-representation*. For example if we use eigenkets of *position operator* as basis, we shall get coordinate representation.

$$\psi_\alpha(\xi) \;\equiv\; \langle\xi|\alpha\rangle. \tag{2.1}$$

(i) The Energy Representation, called the E-representatrion

We consider this as an example of discrete representation. In order to represent the state vector $|\alpha\rangle$ we choose the eigenfunctions of a Hamiltonian operator having a discrete spectrum of eigenvalues as the basis functions. We denote these functions in coordinate representation by

$$
\begin{aligned}
\phi_{E_n}(\xi) &\equiv \langle\xi|E_n\rangle, \\
\phi_{E_n}(\xi)^* &= \langle E_n|\xi\rangle,
\end{aligned}
\tag{2.2}
$$

and $\displaystyle\int d\xi\ \phi_{E_m}(\xi)^*\phi_{E_n}(\xi) = \delta_{E_m,E_n}$ $\qquad(2.3)$

i.e. $\displaystyle\int d\xi\ \langle E_m|\xi\rangle\langle\xi|E_n\rangle \equiv \langle E_m|E_n\rangle = \delta_{E_m.E_n}.$

(ii) Operators as Matrices

We consider any general operator $\hat{X}$ and introduce the *Unit Operator* Eq. (1.35):

$$
\begin{aligned}
\hat{X} &= \hat{I}\cdot\hat{X}\cdot\hat{I} = \left(\sum_n |a_n\rangle\langle a_n|\right)\hat{X}\left(\sum_m |a_m\rangle\langle a_m|\right) \tag{2.4} \\
&= \sum_{n,m} |a_n\rangle\langle a_n|\hat{X}|a_m\rangle\langle a_m| \tag{2.5}
\end{aligned}
$$

The quantities $\langle a_n|\hat{X}|a_m\rangle$ are numbers forming a square array, $\langle a_n|$ are rows and $|a_m\rangle$ the columns. We also know from Eq. (1.25) that

$$
\begin{aligned}
\langle a_n|\hat{X}|a_m\rangle &= \langle a_m|\hat{X}^\dagger|a_n\rangle^*, \qquad \text{and thus} \tag{2.6} \\
\langle a_n|\hat{X}|a_m\rangle^* &= \langle a_m|\hat{X}^\dagger|a_n\rangle, \tag{2.7}
\end{aligned}
$$

in conformity with the definition of Hermitian Adjoint matrix elements as the complex conjugate of the transposed matrix elements.

We also have

$$
\langle a_m|\hat{A}|a_n\rangle = a_n\langle a_n|a_m\rangle = a_n\delta_{n,m}. \tag{2.8}
$$

Thus $\hat{A}$ is diagonal in its own representation.

We can also write

$$
\begin{aligned}
\hat{A} &= \sum_{n,m} |a_n\rangle\langle a_n|\hat{A}|a_m\rangle\langle a_m| \\
&= \sum_m a_m|a_m\rangle\langle a_m| = \sum_m a_m\hat{\Lambda}_m, \tag{2.9}
\end{aligned}
$$

$\hat{\Lambda}_m = |a_m\rangle\langle a_m|$ being the projection operator.

KET AND BRA MATRICES

Let $|\gamma\rangle = \hat{X}|\alpha\rangle$. Then

$$
\begin{aligned}
\langle a_m|\gamma\rangle &= \langle a_m|\hat{X}|\alpha\rangle \\
&= \langle a_m|\hat{X}\left(\sum_n |a_n\rangle\langle a_n|\right)|\alpha\rangle = \sum_n \langle a_m|\hat{X}|a_n\rangle\langle a_n|\alpha\rangle, \quad (2.10)
\end{aligned}
$$

which shows that the product of the *square matrix* $\hat{X} = \langle a_m|\hat{X}|a_n\rangle$ with the *column matrix* $\langle a_n|\alpha\rangle$ reproduces the *column matrix* $\langle a_m|\gamma\rangle$.

Similarly

$$
\text{if } \langle\gamma| = \langle\alpha|\hat{X}, \quad (2.11)
$$

$$
\text{then } \langle\gamma|a_n\rangle = \sum_m \langle\alpha|a_m\rangle\langle a_m|\hat{X}|a_n\rangle. \quad (2.12)
$$

In Eq. (2.12) the right hand side is the product of the *row matrix* $\langle\alpha|a_m\rangle$ with the *square matrix* $\langle a_m|\hat{X}|a_n\rangle$ giving the *row matrix* $\langle\gamma|a_n\rangle$.

Thus we have the *column matrix*

$$
\langle a_m|\gamma\rangle \implies \begin{pmatrix} \langle a_1|\gamma\rangle \\ \langle a_2|\gamma\rangle \\ \vdots \end{pmatrix}, \quad (2.13)
$$

the *square matrix*

$$
\langle a_m|\hat{X}|a_n\rangle \implies \begin{pmatrix} \langle a_1|\hat{X}|a_1\rangle & \langle a_1|\hat{X}|a_2\rangle & \cdots \\ \langle a_2|\hat{X}|a_1\rangle & \langle a_2|\hat{X}|a_2\rangle & \cdots \\ \vdots & \vdots & \ddots \end{pmatrix} \quad (2.14)
$$

and the *row matrix*

$$
\begin{aligned}
\langle\gamma|a_m\rangle \implies & \begin{pmatrix} \langle\gamma|a_1\rangle & \langle\gamma|a_2\rangle & \cdots \end{pmatrix} \\
= & \begin{pmatrix} \langle a_1|\gamma\rangle^* & \langle a_2|\gamma\rangle^* & \cdots \end{pmatrix} \quad (2.15)
\end{aligned}
$$

Problem 2.1 Find in the basis $\{|a_1\rangle,\ |a_2\rangle,\ \cdots\}$
(i) the expression for the inner product $\langle\beta|\alpha\rangle$,
(ii) the matrix corresponding to the outer product $|\beta\rangle\langle\alpha|$.

2.2 Change in Representation and Unitary Transformation

We have seen that a particular representation is determined by the complete set of basis kets used to describe the ket space of the system. A change of representation can then be brought about by changing the basis set. For example, if A and B are two *incompatible i.e. non-commuting* Hermitian operators, each having its distinct set of eigenkets, then either of these sets can be chosen as basis states of the ket space. Thus in place of Eq. (1.33) we can also write

$$|\alpha\rangle = \sum_n |b_n\rangle\langle b_n|\alpha\rangle \tag{2.16}$$

where the expansion coefficients $\langle b_n|\alpha\rangle$ form the wavefunctions in $B-$representation. The transformation in representation can be achieved by a *unitary operator* connecting the old set of basis $\{|a_1\rangle, |a_2\rangle, \cdots \}$ with the new set $\{|b_1\rangle, |b_2\rangle, \cdots \}$ as follows

$$\hat{U} = \sum_n |b_n\rangle\langle a_n| \text{ and } \hat{U}^\dagger = \sum_k |a_k\rangle\langle b_k|. \tag{2.17}$$

$\hat{U}$ defined in Eq. (2.17) can be shown to be unitary.

$$
\begin{aligned}
\hat{U}^\dagger\hat{U} &= \left(\sum_k |a_k\rangle\langle b_k|\right)\left(\sum_n |b_n\rangle\langle a_n|\right)\\
&= \sum_{k,n} |a_k\rangle\langle b_k|b_n\rangle\langle a_n|\\
&= \sum_{k,n} \delta_{k,n}|a_k\rangle\langle a_n| = \sum_k |a_k\rangle\langle a_k|\\
&= \hat{I}.
\end{aligned}
\tag{2.18}
$$

The matrix elements of U in terms of the basis $|a_n\rangle$ are

$$
\begin{aligned}
\langle a_k|\hat{U}|a_l\rangle &= \langle a_k|\left(\sum_m |b_m\rangle\langle a_m|\right)|a_l\rangle\\
&= \sum_m \langle a_k|b_m\rangle\langle a_m|a_l\rangle = \sum_m \langle a_k|b_m\rangle\delta_{m,l}\\
&= \langle a_k|b_l\rangle,
\end{aligned}
\tag{2.19}
$$

which is the inner product of the old bra and the new ket.

Also

$$\hat{U}|a_l\rangle = \sum_k |b_k\rangle\langle a_k|a_l\rangle = |b_l\rangle, \quad \text{and} \tag{2.20}$$

$$\hat{U}^\dagger|b_l\rangle = \sum_k |a_k\rangle\langle b_k|b_l\rangle = |a_l\rangle. \tag{2.21}$$

We can now find out how the expansion coefficients $\langle a_n|\alpha\rangle$ of an arbitrary ket $|\alpha\rangle$ transform from one basis to another,

$$|\alpha\rangle = \sum_n |a_n\rangle\langle a_n|\alpha\rangle, \tag{2.22}$$

$$\langle b_k|\alpha\rangle = \sum_n \langle b_k|a_n\rangle\langle a_n|\alpha\rangle. \tag{2.23}$$

From Eq. (2.20) and Eq. (2.23) follows

$$\langle b_k| = \langle a_k|\hat{U}^\dagger, \tag{2.24}$$

$$\langle b_k|\alpha\rangle = \sum_n \langle a_k|\hat{U}^\dagger|a_n\rangle\langle a_n|\alpha\rangle. \tag{2.25}$$

The left hand sides are the expansion coefficients in new basis and Eq. (2.25) shows the transformation from old coefficient $\langle a_n|\alpha\rangle$ to the new $\langle b_k|\alpha\rangle$. Since the expansion coefficients are the wavefunctions, Eq. (2.25) shows the transformation of wavefunction under change of basis.

TRANSFORMATION OF OPERATOR

For a general operator $\hat{X}$, the matrix elements transforms as follows.

$$
\begin{aligned}
\langle b_k|\hat{X}|b_l\rangle &= \langle b_k|\left(\sum_m |a_m\rangle\langle a_m|\right)\hat{X}\left(\sum_n |a_n\rangle\langle a_n|\right)|b_l\rangle \\
&= \sum_{m,n} \langle b_k|a_m\rangle\langle a_m|\hat{X}|a_n\rangle\langle a_n|b_l\rangle \\
&= \sum_{m,n} \langle a_k|\hat{U}^\dagger|a_m\rangle\langle a_m|\hat{X}|a_n\rangle\langle a_n|\hat{U}|a_l\rangle, \tag{2.26}
\end{aligned}
$$

$$\text{or} \quad \hat{X}_{\text{new}} = \hat{U}^\dagger \hat{X}_{\text{old}}\hat{U}, \tag{2.27}$$

which is the similarity transformation of operators under $\hat{U}$.

2.3   Commuting Observables

Observables A and B are commuting if the corresponding operators $\hat{A}$ and $\hat{B}$ commute with each other:

$$\left[\hat{A}, \hat{B}\right] = \hat{A}\hat{B} - \hat{B}\hat{A} = 0. \tag{2.28}$$

Since $\hat{A}$ and $\hat{B}$ are Hermitian operators, they have complete sets of eigenkets $\{|a_n\rangle\}$ and $\{|b_n\rangle\}$ respectively. If $\hat{A}$ and $\hat{B}$ commute with each other, then they have simultaneous eigenkets.

Proof:

$$\begin{aligned}
\langle a_m|\left[\hat{A}, \hat{B}\right]|a_n\rangle &= \langle a_m|\left(\hat{A}\hat{B} - \hat{B}\hat{A}\right)|a_n\rangle \\
&= \langle a_m|\hat{A}\hat{B}|a_n\rangle - \langle a_m|\hat{B}\hat{A}|a_n\rangle \\
&= a_m^* \langle a_m|\hat{B}|a_n\rangle - a_n \langle a_m|\hat{B}|a_n\rangle \\
&= (a_m^* - a_n) \langle a_m|\hat{B}|a_n\rangle \\
&= (a_m - a_n) \langle a_m|\hat{B}|a_n\rangle = 0.
\end{aligned} \tag{2.29}$$

The last line follows from the hermiticity of $\hat{A}$.

For non-degenerate case, *i.e.* for the case when there is only one eigenket corresponding to a particular eigenvalue, we have

$$\begin{aligned}
\text{for } n \neq m \quad \langle a_m|\hat{B}|a_n\rangle &= 0, \\
\text{and we write } \langle a_m|\hat{B}|a_n\rangle &= \delta_{m.n} \langle a_n|\hat{B}|a_n\rangle \\
&= \delta_{m,n}\, b_n, \quad \text{where } b_n = \langle a_n|\hat{B}|a_n\rangle.
\end{aligned} \tag{2.30}$$

Problem 2.2 Show that the operators $\hat{A}$ and $\hat{B}$ will commute if they have common eigenkets.

Since the common eigenket has the eigenvalue a_n for $\hat{A}$ and the eigenvalue b_n for $\hat{B}$ we express this fact by the notation

$$\hat{A}|a_n, b_n\rangle = a_n|a_n, b_n\rangle, \tag{2.31}$$
$$\hat{B}|a_n, b_n\rangle = b_n|a_n, b_n\rangle, \tag{2.32}$$
$$\text{with} \qquad n = 1, 2, \cdots$$

The operators $\hat{A}$ and $\hat{B}$ are called compatible if they have simultaneous eigenkets. We also note that the measurement of $\hat{A}$ and $\hat{B}$ when the system is in one of the common eigenkets will *precisely* yield the respective eigenvalues a_n and b_n.

2.4 Uncertainty Relation

From the previous discussion it is evident that two *observables* cannot simultaneously have definite values unless the corresponding *hermitian* operators commute. A knowledge of their commutation relation, however, allows us to derive an inequality which their dispersions satisfy.

The deviation can be defined as

$$\Delta \hat{A} = \hat{A} - \langle \hat{A} \rangle. \tag{2.33}$$

Since the expectation value of $\Delta \hat{A}$, $\langle \Delta \hat{A} \rangle = 0$, we *define* the *dispersion of* $\hat{A}$ by the quantity

$$\left\langle \left(\Delta \hat{A} \right)^2 \right\rangle = \left\langle \left(\hat{A} - \langle \hat{A} \rangle \right)^2 \right\rangle$$
$$= \langle \hat{A}^2 \rangle - \langle \hat{A} \rangle^2. \tag{2.34}$$

The dispersion of $\hat{A}$ is zero in its eigenstate. The *uncertainty* of $\hat{A}$ is defined by the square root of the *dispersion*, which is also the *mean square of the deviation* of $\hat{A}$.

If the commutator of two Hermitian operators $\hat{A}$ and $\hat{B}$ is given by

$$\left[\hat{A}, \hat{B} \right] = i\hat{F}, \tag{2.35}$$

it can be shown that

$$\left\langle \left(\Delta \hat{A} \right)^2 \right\rangle \left\langle \left(\Delta \hat{B} \right) \right\rangle^2 \geq \frac{1}{4} \left| \langle \hat{F} \rangle \right|^2 = \frac{1}{4} \left| \left\langle \left[\hat{A}, \hat{B} \right] \right\rangle \right|^2 \tag{2.36}$$

Problem 2.3 Check that $\hat{F}$ in Eq. (2.35) is Hermitian.

Proof of Eq. (2.36).

We consider the expectation values of $\hat{A}$ and $\hat{B}$ in an arbitrary state $|\gamma\rangle$

$$\langle \hat{A} \rangle \equiv \langle \gamma | \hat{A} | \gamma \rangle, \quad \langle \hat{B} \rangle \equiv \langle \gamma | \hat{B} | \gamma \rangle, \tag{2.37}$$

and also define the hermitian operators

$$\Delta \hat{A} = \hat{A} - \langle \hat{A} \rangle \text{ and } \Delta \hat{B} = \hat{B} - \langle \hat{B} \rangle \tag{2.38}$$

Then $\left[\Delta \hat{A}, \Delta \hat{B} \right] = i\hat{F}$. We now construct a ket

$$|\eta\rangle = \left(c\Delta \hat{A} - i\Delta \hat{B} \right) |\gamma\rangle, \tag{2.39}$$

where c is an arbitrary *real* parameter.

Since the *norm* of $|\eta\rangle$ has to be *positive definite* so

$$\langle\gamma| \left(c\Delta\hat{A} + i\Delta\hat{B}\right)\left(c\Delta\hat{A} - i\Delta\hat{B}\right)|\gamma\rangle \; \geq \; 0, \tag{2.40}$$

as c is *real* ($c = c^*$) and $\Delta\hat{A}$ and $\Delta\hat{B}$ are *Hermitian*. Now

$$\left(c\Delta\hat{A} + i\Delta\hat{B}\right)\left(c\Delta\hat{A} - i\Delta\hat{B}\right) \; = \; c^2\left(\Delta\hat{A}\right)^2 + c\hat{F} + \left(\Delta\hat{B}\right)^2 \tag{2.41}$$

Taking the *expectation value* of Eq. (2.41) in the state $|\gamma\rangle$ we finally get

$$\left\langle\left(\Delta\hat{A}\right)^2\right\rangle\left[c + \frac{\langle\hat{F}\rangle}{2\left\langle\left(\Delta\hat{A}\right)^2\right\rangle}\right]^2 + \left\langle\left(\Delta\hat{B}\right)^2\right\rangle - \frac{\langle\hat{F}\rangle^2}{4\left\langle\left(\Delta\hat{A}\right)^2\right\rangle} \; \geq \; 0. \tag{2.42}$$

This is true for any value of c, in particular, for $c = -\dfrac{\langle\hat{F}\rangle}{2\langle(\Delta\hat{A})^2\rangle}$.

Then from Eq. (2.42), we get

$$\left\langle\left(\Delta\hat{B}\right)^2\right\rangle - \frac{\langle\hat{F}\rangle^2}{4\left\langle\left(\Delta\hat{A}\right)^2\right\rangle} \; \geq \; 0, \quad \text{or}$$

$$\left\langle\left(\Delta\hat{A}\right)^2\right\rangle\left\langle\left(\Delta\hat{B}\right)^2\right\rangle \; \geq \; \frac{\langle\hat{F}\rangle^2}{4} \; = \; \frac{1}{4}\left|\left\langle\left[\hat{A}, \hat{B}\right]\right\rangle\right|^2. \tag{2.43}$$

This is the famous *Uncertainty Relation* between two *non-commuting observables*.

We note that in this derivation $\Delta\hat{A}$ and $\Delta\hat{B}$ are operators not *numbers*. When $\hat{A}$ and $\hat{B}$ are *coordinate* and *canonical momentum*, the corresponding uncertainty relation is Heisenberg's *Uncertainty Principle*

Chapter 3

Position & Momentum Operators and Wavefunctions

3.1 Position Operator and its Eigenkets

We define the position operator $\hat{x}$ in one dimension by the eigenvalue equation

$$\hat{x}|x'\rangle \;=\; x'|x'\rangle \tag{3.1}$$

$\hat{x}$ is the operator corresponding to the x-component of position coordinate of the particle, x' is the eigenvalue which is a number having the dimension of length. This is the case of continuous spectrum of eigenvalues where x' may have values from $-\infty$ to $+\infty$. Since position is an observable $\hat{x}$ is Hermitian, and the set of eigenkets $\{|x'\rangle\}$ corresponding to the continuous spectrum of eigenvalues x' form a complete set and can be used as the basis vectors of the ket space. The completeness condition is given by[1]

$$\int_{-\infty}^{+\infty} |x'\rangle\langle x'|\ dx' \;=\; \hat{I}. \tag{3.2}$$

Any arbitrary ket $|\alpha\rangle$ can be expressed in terms of this set:

$$|\alpha\rangle \;=\; \int_{-\infty}^{+\infty} dx'\ |x'\rangle\langle x'|\alpha\rangle. \tag{3.3}$$

The expansion coefficient $\langle x'|\alpha\rangle$ is the wavefunction $\psi_\alpha\left(x'\right)$ in coordinate representation.

---

[1] Compare with Chapter 1 Eq. (1.35) in discrete case.

POSITION PROBABILITY

We have already postulated in the discrete case in Chapter 1 that the probability of obtaining a particular eigenvalue a_n is given by $|\langle a_n|\alpha\rangle|^2$ where $\langle a_n|\alpha\rangle$ is the coefficient of expansion of $|\alpha\rangle$ in the discrete basis $\{|a_n\rangle\}$ of $\hat{A}$. Thus $\langle x'|\alpha\rangle$ in Eq. (3.3) can be defined as the probability amplitude of the position measurement when the system is in the state $|\alpha\rangle$ and $|\langle x'|\alpha\rangle|^2\, dx'$ is the probability of obtaining the value of the x-coordinate between x' and $x' + dx'$, when measurement of x is performed in the state $|\alpha\rangle$ and the detector is placed between x' and $x' + dx'$. The total probability is given by

$$\int_{-\infty}^{+\infty} dx' |\langle x'|\alpha\rangle|^2 \; = \; 1, \tag{3.4}$$

which can also be written as

$$\int_{-\infty}^{+\infty} dx' |\psi_\alpha(x')|^2 \; = \; 1, \tag{3.5}$$

which is the normalization of state vector $|\alpha\rangle$.

We infer that the normalization of the wavefunction is necessary for *probabilistic interpretation of wavefunction* in wave mechanics.

Problem 3.1 Find the dimension of the $\psi_\alpha(x')$ in Eq. (3.5).

INNER PRODUCT

$$\langle\alpha|\beta\rangle \; = \; \int_{-\infty}^{+\infty} \langle\alpha|x'\rangle\langle x'|\beta\rangle dx' \tag{3.6}$$

$$= \; \int_{-\infty}^{+\infty} dx' \psi_\alpha^*(x')\,\psi_\beta(x'), \tag{3.7}$$

which is the overlap between $\psi_\alpha(x')$ and $\psi_\beta(x')$. This can be interpreted as the probability amplitude of state $|\beta\rangle$ to be found in $|\alpha\rangle$ and is independent of representation.[2]

---

[2]Compare the scalar product $\mathbf{a}\cdot\mathbf{b}$ which is the projection of $\mathbf{b}$ along $\mathbf{a}$.

MATRIX ELEMENT $\langle \alpha | \hat{A} | \beta \rangle$

$$
\begin{aligned}
\langle \alpha | \hat{A} | \beta \rangle &= \int dx' \int dx'' \langle \alpha | x' \rangle \langle x' | \hat{A} | x'' \rangle \langle x'' | \beta \rangle \\
&= \int dx' dx'' \psi_\alpha^* (x') \langle x' | \hat{A} | x'' \rangle \psi_\beta (x'') ,
\end{aligned}
\tag{3.8}
$$

where $\langle x' | \hat{A} | x'' \rangle$ is a function of x' and x''.

When

$$
\hat{A} = f(\hat{x}) \tag{3.9}
$$
$$
\langle x' | \hat{A} | x'' \rangle = \langle x' | f(\hat{x}) | x'' \rangle = f(x'') \delta(x' - x'') \tag{3.10}
$$

POSITION EIGENKET IN THREE DIMENSIONS.

We can extend the previous formalism to three dimensions as follows.

$$
\hat{\mathbf{x}} | \mathbf{x}' \rangle = \mathbf{x}' | \mathbf{x}' \rangle, \tag{3.11}
$$

which is equivalent to the following equations

$$
| \mathbf{x}' \rangle = | x_1', x_2', x_3' \rangle, \tag{3.12}
$$
$$
\hat{\mathbf{x}} \longrightarrow (\hat{x}_1, \hat{x}_2, \hat{x}_3) \implies (\hat{x}, \hat{y}, \hat{z})
$$
$$
\left.
\begin{array}{c}
\hat{x}_1 | \mathbf{x}' \rangle \\
\hat{x}_2 | \mathbf{x}' \rangle \\
\hat{x}_3 | \mathbf{x}' \rangle
\end{array}
\right\}
=
\left\{
\begin{array}{c}
\hat{x} | x', y', z' \rangle = x' | x', y', z' \rangle \\
\hat{y} | x', y', z' \rangle = y' | x', y', z' \rangle \\
\hat{z} | x', y', z' \rangle = z' | x', y', z' \rangle
\end{array}
\right.
\tag{3.13}
$$

where x_1, x_2, x_3 are the Cartesian components of the position vector $\mathbf{r}$ and (x_1', x_2', x_3') are the eigenvalues x', y', z'. We note that $| x_1', x_2', x_3' \rangle$ is the simultaneous eigenket of the position coordinates $\hat{x}, \hat{y}, \hat{z}$. Thus $\hat{x}, \hat{y}, \hat{z}$ must mutually commute with each other, *i.e*

$$
[\hat{x}_i, \hat{x}_j] = 0, \text{for all } i, j = 1, 2, 3. \tag{3.14}
$$

We also assume that $\{|\mathbf{x}'\rangle\}$ are complete *i.e.*

$$
\int d^3\mathbf{x}' | \mathbf{x}' \rangle \langle \mathbf{x}' | = \hat{I}. \tag{3.15}
$$

Also, any arbitrary ket $|\alpha\rangle$ can be expanded in term of the complete set Eq. (3.15), *i.e.*

$$|\alpha\rangle = \int d^3x' |\mathbf{x}'\rangle\langle\mathbf{x}'|\alpha\rangle \tag{3.16}$$

$$= \int dx'dy'dz' \, |x'y'z'\rangle\langle x'y'z'|\alpha\rangle.$$

The expansion coefficients $\langle x'y'z'|\alpha\rangle = \langle\mathbf{r}'|\alpha\rangle = \psi_\alpha(\mathbf{r}')$ is the wavefunction in coordinate representation.

3.2   Spatial Translation and Momentum Operator

We will now find out how the position eigenkets $|\mathbf{x}'\rangle$ behave when we spatially translate the system. First we consider an infinitesimal translation $d\mathbf{x}'$ which the position eigenvalue $\mathbf{x}'$ undergoes, so that

$$\mathbf{x}' \longrightarrow \mathbf{x}' + d\mathbf{x}'.$$

This can be achieved either by translating the physical system itself by an amount $d\mathbf{x}'$ or shifting the origin of the coordinate system in the opposite direction, *i.e.* by $-d\mathbf{x}'$. We shall stick to the first approach, known as *active translation*. For this we introduce a *translation operator* $\hat{\mathcal{T}}(d\mathbf{x}')$ defined as follows

$$\hat{\mathcal{T}}(d\mathbf{x}')|\mathbf{x}'\rangle = |\mathbf{x}' + d\mathbf{x}'\rangle. \tag{3.17}$$

$|\mathbf{x}' + d\mathbf{x}'\rangle$ is not an eigenket of $\hat{\mathcal{T}}(d\mathbf{x}')$ but is an eigenket of the position operator $\hat{\mathbf{x}}$ so that

$$\hat{\mathbf{x}}|\mathbf{x}' + d\mathbf{x}'\rangle = (\mathbf{x}' + d\mathbf{x}')|\mathbf{x}' + d\mathbf{x}'\rangle. \tag{3.18}$$

The effect of the operator $\hat{\mathcal{T}}(d\mathbf{x}')$ on an arbitrary state vector $|\alpha\rangle$ can be obtained in the coordinate representation of $|\alpha\rangle$:

$$|\alpha\rangle_{\text{tr}} \equiv \hat{\mathcal{T}}(d\mathbf{x}')|\alpha\rangle = \hat{\mathcal{T}}(d\mathbf{x}')\int d^3x' |\mathbf{x}'\rangle\langle\mathbf{x}'|\alpha\rangle$$

$$= \int d^3x' |\mathbf{x}' + d\mathbf{x}'\rangle\langle\mathbf{x}'|\alpha\rangle, \tag{3.19}$$

where we have used [3]

$$\int d^3x' |\mathbf{x}'\rangle\langle\mathbf{x}'| = \hat{I}.$$

The spatial translation operator $\hat{\mathcal{T}}(d\mathbf{x}')$ should have the following properties

---

[3] $\langle\mathbf{x}'|\alpha\rangle$ being a number is not affected by $\hat{\mathcal{T}}(d\mathbf{x}')$.

(i). It has to be unitary so that the norm of $|\alpha\rangle$ remains invariant and the probability is conserved.

$$\langle \alpha | \alpha \rangle_{\text{tr}} = \langle \alpha | \hat{\mathcal{T}}^{\dagger} (d\mathbf{x}') \hat{\mathcal{T}} (d\mathbf{x}') | \alpha \rangle = \langle \alpha | \alpha \rangle$$
$$\implies \hat{\mathcal{T}}^{\dagger} (d\mathbf{x}') \hat{\mathcal{T}} (d\mathbf{x}') = \hat{I}. \tag{3.20}$$

(ii). The effect of two successive translations by amounts $d\mathbf{x}'$ and $d\mathbf{x}''$ should be a single translation $d\mathbf{x}' + d\mathbf{x}''$ i.e.

$$\hat{\mathcal{T}} (d\mathbf{x}') \hat{\mathcal{T}} (d\mathbf{x}'') = \hat{\mathcal{T}} (d\mathbf{x}' + d\mathbf{x}''). \tag{3.21}$$

(iii). Since

$$\lim_{d\mathbf{x}' \to 0} \hat{\mathcal{T}} (d\mathbf{x}') |\mathbf{x}'\rangle = \lim_{d\mathbf{x}' \to 0} |\mathbf{x}' + d\mathbf{x}'\rangle = |\mathbf{x}'\rangle, \quad \text{so}$$

$$\lim_{d\mathbf{x}' \to 0} \hat{\mathcal{T}} (d\mathbf{x}') = \hat{I}, \quad \text{Identity.} \tag{3.22}$$

(iv). Inverse translation is equivalent to an infinitesimal translation in the opposite direction

$$\hat{\mathcal{T}}^{-1} (d\mathbf{x}') = \hat{\mathcal{T}} (-d\mathbf{x}'), \quad \text{so that} \tag{3.23}$$
$$\hat{\mathcal{T}}^{-1} \hat{\mathcal{T}} = \hat{I} \tag{3.24}$$

Now if we define

$$\hat{\mathcal{T}} (d\mathbf{x}') = 1 - i\hat{\mathbf{k}} \cdot d\mathbf{x}' \tag{3.25}$$

keeping only the first term in the infinitesimal $d\mathbf{x}'$, and demand that $\mathbf{k}$ to be Hermitian operator, then such an expression for $\hat{\mathcal{T}} (d\mathbf{x}')$ satisfy all the four requirements listed above.

Problem 3.2 Show that $\hat{\mathcal{T}} (d\mathbf{x}')$ defined by Eq. (3.25) satisfies the four criteria of the infinitesimal translation operator

Problem 3.3 Prove the following

$$\left[\hat{\mathbf{x}}, \hat{\mathcal{T}} (d\mathbf{x}') \right] |\mathbf{x}'\rangle = d\mathbf{x}' |\mathbf{x}' + d\mathbf{x}'\rangle. \tag{3.26}$$

Now keeping terms up to the first order of smallness in the right hand side of Eq. (3.26) we have

$$\left[\hat{\mathbf{x}}, \hat{\mathcal{T}}\left(d\mathbf{x}'\right)\right]|\mathbf{x}'\rangle \;\approx\; d\mathbf{x}'|\mathbf{x}'\rangle. \tag{3.27}$$

Using the expression for $\hat{\mathcal{T}}\left(d\mathbf{x}'\right)$ in Eq. (3.25), Eq. (3.27) reduces to the following

$$\left[\hat{\mathbf{x}}, -i\hat{\mathbf{k}}\cdot d\mathbf{x}'\right] \;=\; d\mathbf{x}', \qquad \text{or} \tag{3.28}$$

$$-i\hat{\mathbf{x}}\left(\mathbf{k}\cdot d\mathbf{x}'\right) \;+\; i\left(\hat{\mathbf{k}}\cdot d\mathbf{x}'\right)\mathbf{x} \;=\; d\mathbf{x}' \tag{3.29}$$

This is a vector equation. Equating the *i*-th component from both sides of Eq. (3.29)

$$-i\hat{x}_i\sum_j \hat{k}_j dx'_j + i\sum_j \hat{k}_j dx'_j \hat{x}_i \;=\; dx'_i \;=\; \sum_j \delta_{i,j} dx'_j. \tag{3.30}$$

Comparing both sides of Eq. (3.30) we get

$$\hat{x}_i\hat{k}_j - \hat{k}_j\hat{x}_i \;=\; i\delta_{i,j} \quad i.e. \tag{3.31}$$

$$\left[\hat{x}_i, \hat{k}_j\right] \;=\; i\delta_{i,j}. \tag{3.32}$$

We have logically reached a stage when we can identify the Hermitian operator $\hat{\mathbf{k}}$. Since all the terms of left hand side are to be dimensionless from Eq. (3.31), it is obvious that $\hat{\mathbf{k}}$ should have the dimension of inverse of length. Also we know from classical mechanics that translation occurs due to momentum $\mathbf{p}$, we can thus write

$$\hat{\mathbf{k}} \;=\; \frac{\hat{\mathbf{p}}}{\hbar}, \text{ or } \hat{\mathbf{k}}\hbar \;=\; \hat{\mathbf{p}}, \tag{3.33}$$

where $\hbar$ is the fundamental constant having the dimension of *action* and $\hat{\mathbf{k}}$ is inverse of length as stated in Eq. (3.31).

FUNDAMENTAL COMMUTATION RELATION

Putting the expression for $\hat{\mathbf{k}} \;=\; \frac{\hat{\mathbf{p}}}{\hbar}$ in Eq. (3.32) we obtain the *fundamental commutation relations*

$$\hat{x}_m\hat{p}_n \;-\; \hat{p}_n\hat{x}_m \;=\; [\hat{x}_m, \hat{p}_n] \;=\; i\hbar\delta_{m,n} \tag{3.34}$$

The infinitesimal translation operator $\hat{\mathcal{T}}\left(d\mathbf{x}'\right)$ becomes

$$\hat{\mathcal{T}}\left(d\mathbf{x}'\right) \;=\; 1 - i\frac{\hat{\mathbf{p}}\cdot d\mathbf{x}'}{\hbar}, \tag{3.35}$$

where $\hat{\mathbf{p}}$ is identified as the *generator of translation*.

HEISENBERG UNCERTAINTY RELATION

Substitution of Eq. (3.33) in Eq. (2.36) will give the inequality connecting the dispersions of $\hat{x}$ and $\hat{p}_x$

$$\langle (\Delta \hat{x})^2 \rangle \langle (\Delta \hat{p}_x)^2 \rangle \; \geq \; \frac{1}{4}\hbar^2, \tag{3.36}$$

which are the uncertainties Δx and Δp_x. We can write

$$\Delta x \Delta p_x \; \geq \; \hbar/2. \tag{3.37}$$

This is the original Heisenberg Uncertainty Relation between the canonically conjugate variables x and p_x, whereas Eq. (2.36) is the general form connecting dispersions of the operators $\hat{A}$ and $\hat{B}$.

FINITE TRANSLATION

A finite translation can be built up by compounding a very large number of successive infinitesimal translations in the same direction. Using property (ii) of $\hat{\mathcal{T}}(d\mathbf{x}')$ of compounding successive translations, we get a finite translation a in a perticular direction as follows.

$$\hat{\mathcal{T}}(a) \; = \; \lim_{N \to \infty} \left(1 - i\frac{\hat{p}_x}{\hbar}\frac{a}{N} \right)^N \tag{3.38}$$

$$= \; \exp\left(-i\frac{\hat{p}_x a}{\hbar} \right), \tag{3.39}$$

where the exponential form of any operator is to be understood by the infinite series while operating on any function or ket

$$\exp \hat{X} \; = \; 1 + \hat{X} + \frac{1}{1!}\hat{X} + \frac{1}{2!}\hat{X}^2 + \cdots . \tag{3.40}$$

Since successive translations in different directions are commutative, the operators

$$\hat{\mathcal{T}}\left(a\hat{\mathbf{i}} \right) \; = \; \exp\left(-i\frac{a\hat{p}_x}{\hbar} \right) \tag{3.41}$$

$$\hat{\mathcal{T}}\left(b\hat{\mathbf{j}} \right) \; = \; \exp\left(-i\frac{b\hat{p}_y}{\hbar} \right), \tag{3.42}$$

commute with each other, $\hat{\mathbf{i}}$ and $\hat{\mathbf{j}}$ being the unit vectors along x and y directions respectively, a and b being the respective displacements.

Problem 3.4 Obtain the commutator of $\hat{\mathcal{T}}\left(a\hat{\mathbf{i}}\right)$ and $\hat{\mathcal{T}}\left(b\hat{\mathbf{j}}\right)$ and since they commute with each other, show that

$$[\hat{p}_x, \hat{p}_y] \;=\; 0, \quad \text{or more generally } [\hat{p}_i, \hat{p}_j] \;=\; 0. \quad i,j \;=\; x, y, z. \tag{3.43}$$

This is one of the important differences between translation and rotation that unlike translation, rotations about different axes do not commute with each other as we will find out later on. This is manifested in the non-commutability of angular momentum about different axes.

Since $\hat{p}_x$, $\hat{p}_y$, $\hat{p}_z$ are commuting operators they are compatible and have simultaneous eigenkets.

$$|\mathbf{p}'\rangle \;=\; |p'_x, p'_y, p'_z\rangle \tag{3.44}$$

and the eigenvalue equations are

$$\hat{p}_x|\mathbf{p}'\rangle \;=\; \hat{p}_x|p'_x, p'_y, p'_z\rangle \;=\; p'_x|p'_x, p'_y, p'_z\rangle,$$

Similarly for $\hat{p}_y$ and $\hat{p}_z$. Each of the eigenvalues p'_x, p'_y, p'_z have the dimension of momentum. The completeness condition of the eigenkets $|\mathbf{p}'\rangle$ is given by

$$\int d^3\mathbf{p}'|\mathbf{p}'\rangle\langle\mathbf{p}'| \;=\; \hat{I}, \tag{3.45}$$

$$\int dp'_x dp'_y dp'_z \; |p'_x, p'_y, p'_z\rangle\langle p'_x, p'_y, p'_z| \;=\; \hat{I}. \tag{3.46}$$

We also observe that unlike position eigenket $|\mathbf{x}'\rangle$, momentum eigenket $|\mathbf{p}'\rangle$ is also eigenket of $\hat{\mathcal{T}}\left(d\mathbf{x}'\right)$

$$\hat{\mathcal{T}}\left(d\mathbf{x}'\right)|\mathbf{p}'\rangle \;=\; \left(1 - i\frac{\hat{\mathbf{p}} \cdot d\mathbf{x}'}{\hbar}\right)|\mathbf{p}'\rangle$$

$$=\; \left(1 - i\frac{\mathbf{p}' \cdot d\mathbf{x}'}{\hbar}\right)|\mathbf{p}'\rangle \tag{3.47}$$

However, the eigenvalue $\left(1 - i\frac{\mathbf{p}' \cdot d\mathbf{x}'}{\hbar}\right)$ is not real. Since $\hat{\mathcal{T}}\left(d\mathbf{x}'\right)$ is unitary but *not* Hermitian, this is not surprising.

We can now summarize the basic commutation relations

$$[\hat{x}_i, \hat{x}_j] \;=\; 0, \quad [\hat{p}_i, \hat{p}_j] \;=\; 0, \quad [\hat{x}_i, \hat{p}_j] \;=\; i\hbar\delta_{i.j}, \tag{3.48}$$

where i, j are 1, 2, 3 or equivalently x, y, z respectively.

These are called the *canonical commutation relations*. Instead of assuming these relations at the outset, in analogy to the classical *Poisson Bracket* relations (using the *Correspondence Principle*) we have derived them from the properties of unitary infinitesimal translation operator and identifying the generator of translation.

3.3 Momentum Operator in Position Basis

We first consider one-dimensional case, where the position eigenkets $|x'\rangle$ of $\hat{x}$ form a complete set:

$$\int dx' \, |x'\rangle\langle x'| \; = \; \hat{I}, \tag{3.49}$$

forming the basis of ket space of the system.

Operaing $\hat{\mathcal{T}}(\Delta x')$ on an arbitrary system ket $|\alpha\rangle$

$$\hat{\mathcal{T}}(\Delta x')\,|\alpha\rangle \; = \; \int dx' \, \hat{\mathcal{T}}(\Delta x')\,|x'\rangle\langle x'|\alpha\rangle \quad \text{or} \tag{3.50}$$

$$\left(1 - i\,\frac{\hat{p}_x \Delta x'}{\hbar}\right)|\alpha\rangle \; = \; \int dx' \, \hat{\mathcal{T}}(\Delta x')\,|x'\rangle\langle x'|\alpha\rangle \tag{3.51}$$

$$= \; \int dx' \, |x' + \Delta x'\rangle\langle x'|\alpha\rangle. \tag{3.52}$$

$$= \; \int dx'' \, |x''\rangle\langle x'' - \Delta x'|\alpha\rangle \tag{3.53}$$

$$= \; \int dx'' \, |x''\rangle \left[\langle x''|\alpha\rangle \; - \; \Delta x'\frac{\partial}{\partial x''}\langle x''|\alpha\rangle\right] \tag{}$$

$$= \; |\alpha\rangle \; - \; \Delta x' \int dx'' \, |x''\rangle\frac{\partial}{\partial x''}\langle x''|\alpha\rangle. \tag{3.54}$$

Eq. (3.53) follows from Eq. (3.52) by substituting $x'' = x' + \Delta x'$ so that $x' = x'' - \Delta x'$ and $dx'' = dx'$. Now comparison of both sides of Eq. (3.54) yields

$$\hat{p}_x|\alpha\rangle \; = \; \int dx''|x''\rangle \left[-i\hbar\frac{\partial}{\partial x''}\right]\langle x''|\alpha\rangle \quad \text{and} \tag{3.55}$$

$$\langle x'|\hat{p}_x|\alpha\rangle \; = \; \int dx'' \, \langle x'|x''\rangle \left[-i\hbar\frac{\partial}{\partial x''}\right]\langle x''|\alpha\rangle \tag{3.56}$$

$$= \; \int dx''\delta\left(x' - x''\right)\left[-i\hbar\frac{\partial}{\partial x''}\right]\langle x''|\alpha\rangle \tag{3.57}$$

$$= \; -i\hbar\frac{\partial}{\partial x'}\langle x'|\alpha\rangle. \tag{3.58}$$

We also have

$$\langle x'|\hat{p}_x|\alpha\rangle \; = \; \int dx'' \, [\langle x'|\hat{p}_x|x''\rangle]\,\langle x''|\alpha\rangle \tag{3.59}$$

$$= \; \int dx'' \left[-i\hbar\frac{\partial}{\partial x'}\langle x'|x''\rangle\right]\langle x''|\alpha\rangle \tag{3.60}$$

$$= \; -i\hbar\frac{\partial}{\partial x'}\int dx'' \, \delta\left(x' - x''\right)\langle x''|\alpha\rangle. \tag{3.61}$$

Comparing Eq. (3.59) and Eq. (3.61) we obtain

$$\langle x'|\hat{p}_x|x''\rangle = -i\hbar\frac{\partial}{\partial x'}\delta\left(x'-x''\right). \tag{3.62}$$

Using Eq. (3.55)

$$\langle\beta|\hat{p}_x|\alpha\rangle = \int dx' \, \langle\beta|x'\rangle\left[-i\hbar\frac{\partial}{\partial x'}\langle x'|\alpha\rangle\right] \tag{3.63}$$

$$= \int dx' \, \psi_\beta^*\left(x'\right)\left[-i\hbar\frac{\partial}{\partial x'}\right]\psi_\alpha\left(x'\right). \tag{3.64}$$

This form of the operator for $\hat{p}_x$ is derived from the basic properties of momentum as generator of translation in the unitary operator of $\hat{\mathcal{T}}\left(\Delta x\right)$ and is not a postulate.

We can obtain the matrix elements of the nth power of $\hat{p}_x$ by repeated application of the operator:

$$\langle x'|\hat{p}_x^n|\alpha\rangle = \left(-i\hbar\frac{\partial}{\partial x'}\right)^n\langle x'|\alpha\rangle \quad \text{and} \tag{3.65}$$

$$\langle\beta|\hat{p}_x^n|\alpha\rangle = \int dx' \, \psi_\beta^*\left(x'\right)\left(-i\hbar\frac{\partial}{\partial x'}\right)^n\psi_\alpha\left(x'\right). \tag{3.66}$$

3.4   Momentum Wavefunction

The eigenvalue equation of $\hat{p}_x$ is given by

$$\hat{p}_x|p'\rangle = p'|p'\rangle, \tag{3.67}$$

where the eigenkets form an orthonormal complete set

$$\int dp\prime \, |p'\rangle\langle p'| = \hat{I}. \tag{3.68}$$

The orthonormality is given by

$$\langle p'|p''\rangle = \delta\left(p'-p''\right), \qquad \text{and} \tag{3.69}$$

$$|\alpha\rangle = \int dp' \, |p'\rangle\langle p'|\alpha\rangle. \tag{3.70}$$

The expansion coefficients $\langle p'|\alpha\rangle$ can be interpreted as the probability amplitude and $|\langle p'|\alpha\rangle|^2 dp'$ is the probability that a measurement of $\hat{p}_x$ in a state $|\alpha\rangle$ will yield a value between p' and $p'+dp'$ and the *momentum wavefunction* is

$$\langle p'|\alpha\rangle \equiv \phi_\alpha\left(p'\right). \tag{3.71}$$

For a normalized $|\alpha\rangle$,

$$
\begin{aligned}
\langle\alpha|\alpha\rangle &= \int dp' \, \langle\alpha|p'\rangle\langle p'|\alpha\rangle = \int dp' \phi_\alpha^*\left(p'\right)\phi_\alpha\left(p'\right) \\
&= \int dp' \, |\phi_\alpha\left(p'\right)|^2 = 1.
\end{aligned}
\tag{3.72}
$$

The connection between x-representation and p-representation can be obtained using the unitary transformation function. Unitary matrix for a transformation of basis from A representation to B representation is given in Eq. (2.19): $\langle a_k|U|a_l\rangle = \langle a_k|b_l\rangle$. From Eq. (3.58) putting $|\alpha\rangle = |p'\rangle$

$$
\langle x'|\hat{p}_x|p'\rangle = -i\hbar\frac{\partial}{\partial x'}\langle x'|p'\rangle, \quad i.e.
\tag{3.73}
$$

$$
-i\hbar\frac{\partial}{\partial x'}\langle x'|p'\rangle = p'\langle x'|p'\rangle.
\tag{3.74}
$$

The solution of Eq. (3.74) for $\langle x'|p'\rangle$ is

$$
\langle x'|p'\rangle = N\cdot\exp\left(i\frac{p'x'}{\hbar}\right),
\tag{3.75}
$$

where N is the Normalization constant. Thus the momentum eigenfunction in coordinate representation is a plane wave, which is also the solution of the free particle wavefunction.

NORMALIZATION N

we have

$$
\begin{aligned}
\langle x'|x''\rangle &= \int dp' \, \langle x'|p'\rangle\langle p'|x''\rangle \\
&= |N|^2 \int dp' \, \exp\left(i\frac{p'\cdot\left(x'-x''\right)}{\hbar}\right) \\
&= 2\pi\hbar|N|^2\delta\left(x'-x''\right).
\end{aligned}
\tag{3.76}
$$

The left hand side of Eq. (3.76) is $\delta\left(x'-x''\right)$, hence $|N|^2 = \frac{1}{2\pi\hbar}$, where N is chosen as purely real and positive by convention. So

$$
\langle x'|p'\rangle = \frac{1}{\sqrt{2\pi\hbar}}\exp\left(i\frac{p'x'}{\hbar}\right).
\tag{3.77}
$$

Problem 3.5 Show that the position wavefunction $\psi_\alpha(x')$ and the momentum wavefunction $\phi_\alpha(p')$ are Fourier Transforms[4] of each other:

$$\psi_\alpha(x') = \frac{1}{\sqrt{2\pi\hbar}} \int dp' \, \phi_\alpha(p') \exp\left(i\frac{p'x'}{\hbar}\right), \tag{3.78}$$

$$\phi_\alpha(p') = \frac{1}{\sqrt{2\pi\hbar}} \int dx' \, \psi_\alpha(x') \exp\left(-i\frac{p'x'}{\hbar}\right). \tag{3.79}$$

Problem 3.6 The wavefunction of a particle is given by

$$\psi(x) = \phi(x)\exp\left(\frac{ip_0 x}{\hbar}\right), \quad \phi(x) \text{ being a real function.}$$

What is the physical meaning of the quantity p_0?
[Hint: Find $\langle \psi(x) | \hat{p}_x | \psi(x) \rangle$.]

3.5 Gaussian as Minimum Uncertainty Wave Packet

We have

$$[\hat{x}, \hat{p}_x] = i\hbar, \qquad \text{and} \tag{3.80}$$

$$\Delta x \Delta p_x \geq \frac{1}{2}\hbar, \qquad \text{where} \tag{3.81}$$

$$\Delta x = \left[\langle(\hat{x} - \langle\hat{x}\rangle)^2\rangle\right]^{\frac{1}{2}}, \qquad \text{and} \tag{3.82}$$

$$\Delta p_x = \left[\langle(\hat{p}_x - \langle\hat{p}_x\rangle)^2\rangle\right]^{\frac{1}{2}} \tag{3.83}$$

are the uncertainties of $\hat{x}$ and $\hat{p}_x$ in a state $|\psi\rangle$. We recapitulate Eq. (2.40)

$$\langle\gamma|\left(c\Delta\hat{A} + i\Delta\hat{B}\right)\left(c\Delta\hat{A} - i\Delta\hat{B}\right)|\gamma\rangle \geq 0, \qquad \text{for}$$

$\hat{A} = \hat{x}, \ \hat{B} = \hat{p}_x, \ \Delta\hat{A} = \hat{x} - \langle\hat{x}\rangle, \ \Delta\hat{B} = \hat{p}_x - \langle\hat{p}_x\rangle.$

We also have Eq. (2.42)

$$\langle\left(\Delta\hat{A}\right)^2\rangle\left[c + \frac{\langle\hat{F}\rangle}{2\langle\left(\Delta\hat{A}\right)^2\rangle}\right]^2 + \langle\left(\Delta\hat{B}\right)^2\rangle - \frac{\langle\hat{F}\rangle^2}{4\langle\left(\Delta\hat{A}\right)^2\rangle} \geq 0. \tag{3.84}$$

---

[4]See Appendix A.1.

The wavefunction $|\psi\rangle$ that satisfies the equality in Eq. (2.40) with the $c = c_0$ given by

$$c_0 = -\frac{\langle\hat{F}\rangle}{2\langle(\Delta\hat{A})^2\rangle} \tag{3.85}$$

gives the minimum uncertainty product of $\hat{x}$ and $\hat{p}_x$. Thus $|\psi\rangle$ is the solution of the equation

$$(c_0\Delta\hat{x} - i\Delta\hat{p}_x)|\psi\rangle = 0, \quad \text{or} \tag{3.86}$$

$$\left[-\frac{\hbar(x - \langle\hat{x}\rangle)}{2\langle(x - \langle x\rangle)^2\rangle} - i(\hat{p}_x - \langle\hat{p}_x\rangle)\right]|\psi\rangle = 0. \tag{3.87}$$

In coordinate representation Eq. (3.87) becomes

$$-i\hbar\frac{d}{dx}\psi(x) = \left\{\langle p_x\rangle + \frac{i\hbar}{2\langle(\Delta x)^2\rangle}[x - \langle x\rangle]\right\}\psi(x). \tag{3.88}$$

On integration

$$\psi(x) = C\exp\left(\frac{i}{\hbar}\langle\hat{p}_x\rangle x\right)\exp\left(-\frac{(x - \langle\hat{x}\rangle)^2}{4\langle(\Delta x)^2\rangle}\right). \tag{3.89}$$

Introducing the wavenumber $k = \langle\hat{p}\rangle/\hbar$, centering at the origin (*i.e.* $\langle\hat{x}\rangle = 0$) and using the symbol $d^2 = 2\langle(\Delta x)^2\rangle$ we get the Gaussian as the *minimum uncertainty wave packet* with the width d

$$\psi(x) = \frac{1}{\sqrt{\pi}d}\exp\left[-ikx - \frac{x^2}{2d^2}\right]. \tag{3.90}$$

Problem 3.7

(i). Check the normalization in Eq. (3.90).

(ii). Compute the expectation values of $\hat{x}$, $\hat{x}^2$, $\hat{p}_x$, $\hat{p}_x^2$ and verify the following

$$\langle x\rangle = 0, \ \langle x^2\rangle = \frac{1}{2}d^2, \ \langle p_x\rangle = k\hbar, \quad \text{and} \tag{3.91}$$

$$\langle p_x^2\rangle = \frac{\hbar^2}{2d^2} + k^2\hbar^2. \tag{3.92}$$

Problem 3.8 Obtain the wavefunction Eq. (3.89) in momentum representation. You may use $\langle p'|\alpha\rangle = \int dx' \langle p'|x'\rangle\langle x'|\alpha\rangle$.

3.6   Extension to Three Dimension

We have

$$\hat{\mathbf{p}}|\mathbf{p}'\rangle = \mathbf{p}'|\mathbf{p}'\rangle. \tag{3.93}$$

Orthonormality of $|\mathbf{x}'\rangle$ are given by

$$
\begin{aligned}
\langle \mathbf{x}'|\mathbf{x}''\rangle &= \delta\left(\mathbf{x}' - \mathbf{x}''\right) = \delta\left(x' - x''\right)\delta\left(y' - y''\right)\delta\left(z' - z''\right), \tag{3.94}\\
\langle \mathbf{p}'|\mathbf{p}''\rangle &= \delta\left(\mathbf{p}' - \mathbf{p}''\right) = \delta\left(p_x' - p_x''\right)\delta\left(p_y' - p_y''\right)\delta\left(p_z' - p_z''\right). \tag{3.95}
\end{aligned}
$$

Also we have

$$\int d^3\mathbf{x}'\, |\mathbf{x}'\rangle\langle\mathbf{x}'| = \hat{I} \tag{3.96}$$

$$\int d^3\mathbf{p}'\, |\mathbf{p}'\rangle\langle\mathbf{p}'| = \hat{I} \tag{3.97}$$

$$|\alpha\rangle = \int d^3\mathbf{x}'\, |\mathbf{x}'\rangle\langle\mathbf{x}'|\alpha\rangle, \qquad \text{and} \tag{3.98}$$

$$|\alpha\rangle = \int d^3\mathbf{p}'\, |\mathbf{p}'\rangle\langle\mathbf{p}'|\alpha\rangle, \tag{3.99}$$

The expansion coefficients $\langle\mathbf{x}'|\alpha\rangle$ and $\langle\mathbf{p}'|\alpha\rangle$ are the respective wavefunctions in position and momentum spaces.

Chapter 4

Time Evolution of Quantum Systems

4.1 Time Evolution Operator

Our ultimate objective is to know the dynamics of quantum system and obtain the equation of motion. We emphasize that time in non-relativistic mechanics is not a dynamical variable and we have treated it like a parameter.

We introduce the time evolution operator $\hat{U}(t, t_0)$ which takes the state $|\alpha, t_0\rangle$ at a particular time t_0 to a later time state $|\alpha, t_0; t\rangle$ as follows

$$|\alpha, t_0; t\rangle = \hat{U}(t, t_0) |\alpha, t_0\rangle, \quad \text{where} \quad t > t_0. \tag{4.1}$$

This is consistent with the principle of causality in quantum mechanics. The operator $\hat{U}$ in Eq. (4.1) should satisfy the following:

(i). $\hat{U}$ must be unitary. This is to conserve normalization. Thus

$$\langle \alpha, t_0; t | \alpha, t_0; t \rangle = \langle \alpha, t_0 | \alpha, t_0 \rangle, \tag{4.2}$$

and we have

$$\langle \alpha, t_0 | \hat{U}^\dagger(t, t_0) \hat{U}(t, t_0) | \alpha, t_0 \rangle = \langle \alpha, t_0 | \alpha, t_0 \rangle, \tag{4.3}$$

$$i.e. \quad \hat{U}^\dagger \hat{U} = \hat{I}. \tag{4.4}$$

(ii). We also require $\hat{U}$ to have the composition property:

$$\hat{U}(t_2, t_0) = \hat{U}(t_2, t_1) \hat{U}(t_1, t_0), \quad t_2 > t_1 > t_0. \tag{4.5}$$

From Eq. (4.1) we have for infinitesimal time evcolution

$$|\alpha, t_0; t_0 + dt\rangle = \hat{U}(t_0 + dt, t_0)|\alpha, t_0\rangle. \tag{4.6}$$

Now since time is a continuous parameter we can take the following limit:

$$\lim_{dt \to 0} |\alpha, t_0; t_0 + dt\rangle = |\alpha, t_0\rangle. \tag{4.7}$$

Thus we get

$$\lim_{dt \to 0} \hat{U}(t_0 + dt; t_0) = \hat{I} \tag{4.8}$$

as the unit operator.

All the above properties of the time evolution operator follow if we define the infinitesimal time evolution operator as

$$\hat{U}(t_0 + dt; t_0) = \hat{I} - i\hat{\Omega}dt, \tag{4.9}$$

where $\hat{\Omega}^\dagger = \hat{\Omega}$ is Hermitian. From Eq. (4.9) we observe that $\hat{\Omega}$ should have the dimension of inverse of time. We use Planck-Einstein relation $E = \hbar\omega$ connecting energy E and angular frequency ω and write

$$\hat{\Omega} = \frac{\hat{H}}{\hbar} \qquad \text{and} \tag{4.10}$$

$$\hat{U}(t_0 + dt, t_0) = \hat{I} - i\frac{\hat{H}}{\hbar}dt. \tag{4.11}$$

$\hat{H}$ is the Hamiltonian of the system. This expression for $\hat{U}$ is consistent with classical notion that the Hamiltonian is the generator of time evolution.

We have already introduced $\hbar$ in the infinitesimal translation operator $\hat{\mathcal{T}}(d\mathbf{x}')$ in Eq. (3.35)

$$\hat{\mathcal{T}}(d\mathbf{x}') = 1 - i\frac{\hat{\mathbf{p}} \cdot d\mathbf{x}'}{\hbar}, $$

In Eq. (4.11) $\hbar$ has to be the same $\hbar$ as in the Eq. (3.35), otherwise the relation like

$$\frac{d\mathbf{x}}{dt} = \frac{\mathbf{p}}{m} \tag{4.12}$$

cannot be obtained as the classical limit of the corresponding quantum mechanical relation. Now in Eq. (4.5) we put $t_1 = t$, $t_2 = t + dt$ and get

$$\hat{U}(t + dt, t_0) = \hat{U}(t + dt, t)\hat{U}(t, t_0), \qquad \text{and also} \tag{4.13}$$

$$\hat{U}(t + dt, t_0) = \left(\hat{I} - i\frac{\hat{H}dt}{\hbar}\right)\hat{U}(t, t_0). \tag{4.14}$$

Since dt is an infinitesimal we finally obtain

$$i\hbar\frac{\partial}{\partial t}\hat{U}\left(t,t_0\right) = \hat{H}\hat{U}\left(t,t_0\right). \tag{4.15}$$

This is the Schrödinger Equation for time evolution operator $\hat{U}$. The solution of Eq. (4.15) depends on the nature of time dependence of the Hamiltonian. If $\hat{H}$ is independent of time, then the solution is

$$\hat{U}\left(t,t_0\right) = \exp\left[-\frac{i}{\hbar}\hat{H}\left(t-t_0\right)\right]. \tag{4.16}$$

Problem 4.1 Expand the exponential operator in an infinite series and check that Eq. (4.16) satisfies Eq. (4.15).

4.2   The Schrödinger Equation of Motion

Operating on the state ket $|\alpha, t_0\rangle$ we obtain from Eq. (4.15)

$$i\hbar\frac{\partial}{\partial t}\hat{U}\left(t,t_0\right)|\alpha,t_0\rangle = \hat{H}\hat{U}\left(t,t_0\right)|\alpha,t_0\rangle. \tag{4.17}$$

Since

$$\hat{U}\left(t,t_0\right)|\alpha,t_0\rangle = |\alpha,t_0;t\rangle, \tag{4.18}$$

we get

$$i\hbar\frac{\partial}{\partial t}|\alpha,t_0;t\rangle = \hat{H}|\alpha,t_0;t\rangle, \quad \text{or putting } t_0 = 0,$$

$$i\hbar\frac{\partial}{\partial t}|\alpha,t\rangle = \hat{H}|\alpha,t\rangle. \tag{4.19}$$

In position representation the Hamiltonian being

$$\hat{H} = \frac{\hat{\mathbf{p}}^2}{2m} + V\left(\mathbf{r},t\right) = -\frac{\hbar^2}{2m}\nabla^2 + V\left(\mathbf{r},t\right),$$

we get

$$i\hbar\frac{\partial}{\partial t}\psi_\alpha\left(\mathbf{r},t\right) = \left[-\frac{\hbar^2}{2m}\nabla^2 + V\left(\mathbf{r},t\right)\right]\psi_\alpha\left(\mathbf{r},t\right), \tag{4.20}$$

where $\psi_\alpha\left(\mathbf{r},t\right)$ is the wavefunction and Eq. (4.20) is the famous *Schrödinger Equation* of motion.

This equation was originally *postulated* by Schrödinger in 1926. He proposed this equation from the notion that there is some kind of canonical analogy between time and energy similar to position and momentum coordinate. Here the equation is derived from the properties of the unitary time evolution operator where time is treated like a parameter and not a dynamical variable.

INTERPRETATION OF THE WAVEFUNCTION

We recapitulate some concepts of wave mechanics in the light of the formalism developed here. Like the state vector $|\alpha\rangle$, $|\alpha, t\rangle$ also is postulated to contain all the information that can be known about the system, the wavefunction $\psi_\alpha(\mathbf{r}, t)$ which is a position representation of $|\alpha, t\rangle$ will provide a quantum mechanically complete description of the dynamical behaviour of a particle. Since $\psi_\alpha(\mathbf{r}, t)$ are the expansion coefficients of $|\alpha, t\rangle$ in position basis,

$$\begin{aligned}
|\alpha, t\rangle &= \int d^3\mathbf{r}'\, |\mathbf{r}'\rangle\langle\mathbf{r}'|\alpha, t\rangle, \\
\langle\mathbf{r}|\alpha, t\rangle &= \psi_\alpha(\mathbf{r}, t),
\end{aligned}$$

where $|\psi_\alpha(\mathbf{r}, t)|^2$ is interpreted as the *position probability density* P:

$$P(\mathbf{r}, t)\, d^3\mathbf{r} = |\psi_\alpha(\mathbf{r}, t)|^2 d^3\mathbf{r}. \tag{4.21}$$

This means that $P(\mathbf{r}, t)\, d^3\mathbf{r}$ is the probability of finding a particle in the volume element $d^3\mathbf{r}$ about $\mathbf{r}$ at time t. Since the particle must be somewhere in the region Ω^1, we should have

$$\int_\Omega |\psi_\alpha(\mathbf{r}, t)|^2 d^3\mathbf{r} = 1, \tag{4.22}$$

which is also the normalization condition. As the normalization should be independent of time

$$\frac{\partial}{\partial t} \int_\Omega P(\mathbf{r}, t)\, d^3\mathbf{r} = 0. \tag{4.23}$$

Now

$$\begin{aligned}
\frac{\partial}{\partial t} \int_\Omega P(\mathbf{r}, t)\, d^3\mathbf{r} &= \int_\Omega \left[\psi^* \frac{\partial\psi}{\partial t} + \frac{\partial\psi^*}{\partial t}\psi \right] d^3\mathbf{r} \\
&= \frac{i\hbar}{2m} \int_\Omega \nabla \cdot [\psi^*(\nabla\psi) - (\nabla\psi^*)\psi]\, d^3\mathbf{r} \\
&= \frac{i\hbar}{2m} \int_S [\psi^*(\nabla\psi) - (\nabla\psi^*)\psi]_n\, d\hat{S}_n, \tag{4.24}
\end{aligned}$$

---

[1]Not to be confused with $\hat{\Omega}$ in Eq. (4.9)

where Eq. (4.20) with a real V is used. Here S is the bounding surface of Ω and $[\cdots]_n$ denotes the component of the vector within the square bracket in the direction of the outward normal to the surface element $d\mathbf{S}$.

Defining the probability flux or current density as

$$\mathbf{j}(\mathbf{r}, t) = \frac{\hbar}{2im} [\psi^* (\nabla \psi) - (\nabla \psi^*) \psi].$$ (4.25)

We have

$$\frac{\partial}{\partial t} \int P d^3 \mathbf{r} = -\int_\Omega \nabla \cdot \mathbf{j} \, d^3 \mathbf{r}$$

$$= -\int_S \hat{j}_n \, dS_n.$$ (4.26)

In the case of a wave packet (which is localized) ψ vanishes at large distance; thus ψ, $\nabla \psi$ vanish on the surface at infinity and the surface integral in Eq. (4.26) vanishes and Eq. (4.23) is satisfied.

Eq. (4.26) can be written as

$$\frac{\partial P}{\partial t} + \nabla \cdot \mathbf{j} = 0,$$ (4.27)

which is known as *Equation of Continuity*.

4.3 Time Dependence of Expectation Values: Ehrenfest Theorem

We consider the time development of the expectation value of $\hat{x}$

$$\frac{d}{dt} \langle \hat{x} \rangle = \frac{d}{dt} \int \psi^* (\mathbf{r}, t) x \psi (\mathbf{r}.t) d^3 \mathbf{r}.$$ (4.28)

Problem 4.2 Using Green's first identity
$\int_\Omega [u (\nabla^2 v) + (\nabla u) \cdot (\nabla v)] d^3 \mathbf{r} = \int_S u (\nabla v) \cdot d\mathbf{S}$
with the boundary condition, u and $v \rightarrow 0$ on the surface at infinity, show that

$$\frac{d}{dt} \langle x \rangle = -\frac{i\hbar}{m} \int_\Omega \psi^* \frac{\partial}{\partial x} \psi \, d^3 \mathbf{r}.$$ (4.29)

So we get

$$\frac{d}{dt} \langle \hat{x} \rangle = \frac{\langle \hat{p}_x \rangle}{m}.$$ (4.30)

Similarly, we have

$$\frac{d}{dt} \langle \hat{p}_x \rangle = \frac{d}{dt} \int_\Omega \psi^* \left(-i\hbar \frac{\partial}{\partial x} \right) \psi \, d^3 \mathbf{r}.$$ (4.31)

Problem 4.3 Use Green's second identity

$$\int_\Omega \left[u \left(\nabla^2 v \right) - v \left(\nabla^2 u \right) \right] \, d^3\mathbf{r} \;=\; \int_S \left[u \left(\nabla v \right) - v \left(\nabla u \right) \right] \cdot d\mathbf{S} \qquad (4.32)$$

and Eq. (4.31) to show that

$$\frac{d}{dt} \langle \hat{p}_x \rangle \;=\; - \left\langle \frac{\partial V}{\partial x} \right\rangle. \qquad (4.33)$$

The equations Eq. (4.30) and Eq. (4.31) are analogous to the classical equations

$$\frac{d\mathbf{r}}{dt} \;=\; \frac{\mathbf{p}}{m} \quad \text{and} \quad \frac{d\mathbf{p}}{dt} \;=\; -\nabla V,$$

and are known as the *Ehrenfest Theorem*.

4.4   The Schrödinger and Heisenberg Pictures

There are many representations of state vectors and observables connected by unitary transformations. We distinguish between two classes of representations which differ in the way the time evolution of the system is achieved. These are called *pictures*. In *Schrödinger Picture* the state vectors evolve in time, whereas operators corresponding to dynamical variables like position and momentum are independent of time.

$$|\alpha\rangle \;\longrightarrow\; |\alpha, t\rangle \;=\; \hat{U}(t)\,|\alpha\rangle, \quad \text{and} \quad \hat{A} \;\longrightarrow\; \hat{A}. \qquad (4.34)$$

Here the initial time t_0 is taken to be 0. Since the inner product remains invariant under unitary transformation

$$\langle \alpha | \beta \rangle \;\longrightarrow\; \langle \alpha | \hat{U}^\dagger \hat{U} | \beta \rangle \;=\; \langle \alpha | \beta \rangle. \qquad (4.35)$$

Also

$$\langle \alpha | \hat{A} | \beta \rangle \;\longrightarrow\; \langle \alpha | \hat{U}^\dagger \hat{A} \hat{U} | \beta \rangle. \qquad (4.36)$$

From Eq. (4.36) it is evident that instead of state vectors transforming under $\hat{U}$, the other approach is operators transforming and state vectors remaining unchanged under $\hat{U}$. This is known as *Heisenberg Picture*, where

$$\hat{A} \;\longrightarrow\; \hat{A}^{(H)}(t) \;=\; \hat{U}^{-1} \hat{A}^{(S)} \hat{U} \quad \text{and} \quad |\alpha\rangle \;\longrightarrow\; |\alpha\rangle. \qquad (4.37)$$

The superscripts (H) and (S) refer to Heisenberg and Schrödinger pictures respectively.

$$\text{At} \quad t = 0 \qquad \hat{A}^{(H)}(t = 0) = \hat{A}^{(S)}, \tag{4.38}$$

and the state vectors in both the pictures coincide at $t = 0$. At a later time t, the state vector in Schrödinger picture evolves in time by $\hat{U}(t)$ according to Eq. (4.34), whereas the state vector in Heisenberg picture remains frozen in time to that at $t = 0$.

$$|\alpha, t_0 = 0; t\rangle_H = |\alpha, t_0 = 0\rangle, \tag{4.39}$$

and

$$|\alpha, t_0; t\rangle_S = \hat{U}(t) |\alpha, t_0 = 0\rangle, \tag{4.40}$$

$$|\alpha, t_0 = 0; t\rangle_H = |\alpha, t_0 = 0\rangle = \hat{U}^\dagger(t) |\alpha, t_0 = 0; t\rangle_S$$

$$= \exp\left[\frac{i}{\hbar}\hat{H}(t - t_0)\right] |\alpha, t_0 = 0; t\rangle_S. \tag{4.41}$$

4.5 The Heisenberg Equation of Motion

Since

$$\hat{A}^{(H)} = \hat{U}^{-1}\hat{A}^{(S)}\hat{U},$$

$$\frac{d}{dt}\hat{A}^{(H)} = \frac{d}{dt}\left[\hat{U}^{-1}\hat{A}^{(S)}\hat{U}\right] \tag{4.42}$$

$$= \left(\frac{\partial \hat{U}^{-1}}{\partial t}\right)\hat{A}^{(S)}\hat{U} + \hat{U}^{-1}\hat{A}^{(S)}\left(\frac{\partial \hat{U}}{\partial t}\right), \tag{4.43}$$

as the operator $\hat{A}$ in the Schrödinger picture is not an explicit function of time. Also from Eq. (4.15)

$$\frac{\partial \hat{U}^{-1}}{\partial t} = \frac{\partial \hat{U}^\dagger}{\partial t} = -\frac{1}{i\hbar}\hat{U}^{-1}\hat{H}, \quad \text{and} \tag{4.44}$$

$$\frac{\partial \hat{U}}{\partial t} = \frac{1}{i\hbar}\hat{H}\hat{U}. \tag{4.45}$$

Therefore

$$\frac{d\hat{A}^{(H)}}{dt} = \frac{1}{i\hbar}\hat{U}^{-1}\left[\hat{A}^{(S)}\hat{H} - \hat{H}\hat{A}^{(S)}\right]\hat{U} \tag{4.46}$$

$$= \frac{1}{i\hbar}\left[\hat{A}^{(H)}, \hat{H}^{(H)}\right]. \tag{4.47}$$

In those cases where $\hat{U}$ is given by Eq. (4.16) then $\hat{H}$ commutes with $\hat{U}$ and

$$\frac{d\hat{A}^{(H)}}{dt} = \frac{1}{i\hbar} \left[\hat{A}^{(H)}, \hat{H} \right]. \tag{4.48}$$

This is the *Heisenberg Equation of Motion*.

4.6   Operator Form of the Hamiltonian: Classical Analogue

For a physical system having a classical analogue, we assume the Hamiltonian to be of the same form as the classical one, with classical position and momentum components x_i and p_i being replaced by the corresponding quantum mechanical operators. If any ambiguity arises in the product operators which are non-commuting, then we use the criterion that the Hamiltonian has to be Hermitian. Thus a classical product xp is replaced by $\frac{1}{2} \left(\hat{x}\hat{p} + \hat{p}\hat{x} \right)$ to obtain the corresponding quantum mechanical operator. While calculating the operator for commutators of x_i and p_i with the functions of x_i and p_i, one may use the following formulas

$$[\hat{x}_i, F(\hat{\mathbf{p}})] = i\hbar \frac{\partial F}{\partial \hat{p}_i}, \quad \text{and} \tag{4.49}$$

$$[\hat{p}_i, G(\hat{\mathbf{x}})] = -i\hbar \frac{\partial G}{\partial \hat{x}_i}. \tag{4.50}$$

Example 4.1 Free Particle.
For a free particle we have

$$\hat{H} = \frac{\mathbf{p}^2}{2m} = \sum_{j=1}^{3} \frac{\hat{p}_j^2}{2m}. \tag{4.51}$$

So

$$\frac{d\hat{p}_i}{dt} = \frac{1}{i\hbar} \left[\hat{p}_i, \hat{H} \right] = 0,$$

$$\hat{p}_i(t) = \hat{p}_i(0), \quad \text{a constant of motion,}$$

$$\frac{d\hat{x}_i}{dt} = \frac{1}{i\hbar} \left[\hat{x}_i, \hat{H} \right] = \frac{1}{i\hbar} \left[\hat{x}_i, \sum_{j=1}^{3} \frac{\hat{p}_j^2}{2m} \right]$$

$$= \frac{1}{i\hbar} \left[\hat{x}_i, \frac{\hat{p}_i^2}{2m} \right] = \frac{\hat{p}_i}{m}$$

$$= \frac{\hat{p}_i(0)}{m}, \quad \text{and thus} \tag{4.52}$$

$$\hat{x}_i(t) = \hat{x}_i(0) + \frac{\hat{p}_i(0)}{m}t. \tag{4.53}$$

Thus equal time commutator of $\hat{x}_i(0)$ and $\hat{x}_j(0)$ is zero

$$[\hat{x}_i(0), \hat{x}_j(0)] = 0. \tag{4.54}$$

However, $\hat{x}_i(t)$ and $\hat{x}_i(0)$ does not commute, because of the presence of $\hat{p}_i(0)$ in $\hat{x}_i(t)$.

Problem 4.4 Show that

$$[\hat{x}_i(t), \hat{x}_i(0)] = \frac{-i\hbar t}{m}, \tag{4.55}$$

and hence from the uncertainty relation Eq. (2.43)

$$\langle(\Delta\hat{x}_i(t))^2\rangle\langle(\Delta\hat{x}_i(0))^2\rangle \geq \frac{\hbar^2 t^2}{4m^2}. \tag{4.56}$$

thus even if at $t = 0$ the particle is well localized, its position becomes more and more uncertain as it spreads with time. This is also demonstrated in the case of wave packet.

4.7 Time Dependence of the Base Kets

If we define the complete set of kets of $\hat{A}$ as the basis ket vectors of the space at time $t = 0$ where

$$\hat{A}|a_n\rangle = a_n|a_n\rangle. \tag{4.57}$$

Since in the Schrödinger picture the operator $\hat{A}$ does not change with time, the eigenvalue equation Eq. (4.57) remains the same as at $t = 0$ and hence $|a_n\rangle$ does not depend on time. Thus unlike the state vectors, base kets do not change with time.

However, in Heisenberg picture the situation is different. Since

$$\hat{A}^{(H)}(t) = \hat{U}^\dagger \hat{A}(t = 0)\hat{U}, \tag{4.58}$$

and from Eq. (4.57) we have

$$\hat{U}^\dagger \hat{A}(0)\hat{U}\hat{U}^\dagger|a_n\rangle = a_n\hat{U}^\dagger|a_n\rangle, \tag{4.59}$$

$$\text{i.e.} \quad \hat{A}^{(H)}\hat{U}^\dagger|a_n\rangle = a_n\hat{U}^\dagger|a_n\rangle, \tag{4.60}$$

which is the eigenvalue equation for $\hat{A}^{(H)}$ with eigenket changing from $|a_n\rangle$ at $t = 0$ to $\hat{U}^\dagger|a_n\rangle$ at time t. Thus we have to use the set $\left\{\hat{U}^\dagger|a_n\rangle\right\}$ as the basis kets in

Heisenberg picture in place of the stationary kets $\{|a_n\rangle\}$ as basis in Schrödinger picture.

We also have

$$\hat{A}^{(H)}(t) = \hat{A}^{(H)}(t) \sum_{a_n} |a_n, t\rangle_{HH}\langle a_n, t| \tag{4.61}$$

$$\text{or } \hat{A}^{(H)}(t) = \sum_{a_n} |a_n, t\rangle a_n \langle a_n, t|$$

$$= \sum_{a_n} \hat{U}^\dagger |a_n\rangle a_n \langle a_n| \hat{U} \tag{4.62}$$

$$= \hat{U}^\dagger \hat{A}^{(S)} \hat{U}, \tag{4.63}$$

which is the same as Eq. (4.37).

Thus we tabulate the properties of the Schrödinger versus the Heisenberg pictures.

Table 4.1 The two types of description of evolution of quantum states.

	Schrödinger Picture	**Heisenberg Picture**
STATE KETS	*Evolve in time*	*Stationary*
OPERATORS	*Stationary*	*Evolve in time*
BASE KETS	*Stationary*	*Evolve oppositely*

Chapter 5

Propagators and Feynman Path Integral

The path integral approach to quatum mechanics was developed by Feynman when he was a graduate student at Princeton University. The method provides a deep insight into quantum dynamics. Due to computational complexity the path integral formalism is not convenient to deal problems in non-relativistic quantum mechanics. It, however, provides an excellent method for quantizing quantum fields and has become a powerful tool in quantum field theory, statistical physics and numerical computation.

5.1 Propagators

In wave mechanics one way of solving the time evolution problem with a time independent Hamiltonian $\hat{H}$ is by expanding the initial state ket $|\alpha\rangle$ in terms of the eigenkets $\{|a_n\rangle\}$ of an observable $\hat{A}$ that commutes with $\hat{H}$. We thus write

$$|\alpha, t_0; t\rangle = \exp\left[-i\frac{\hat{H}(t - t_0)}{\hbar}\right]|\alpha, t_0\rangle$$

$$= \sum_{a_n}|a_n\rangle\langle a_n|\alpha, t_0\rangle \exp\left[-i\frac{E_n(t - t_0)}{\hbar}\right] \quad (5.1)$$

where $\qquad \hat{A}|a_n\rangle = a_n|a_n\rangle$

and $\qquad \sum_{a_n}|a_n\rangle\langle a_n| = \hat{I}.$

Coordinate representation of the above equation is

$$\langle \mathbf{x}' | \alpha, t_0; t \rangle \;=\; \sum_{a_n} \langle \mathbf{x}' | a_n \rangle \langle a_n | \alpha, t_0 \rangle \exp \left[-i \frac{E_n (t - t_0)}{\hbar} \right] \tag{5.2}$$

$$\text{or} \quad \psi(\mathbf{x}', t) \;=\; \sum_{a_n} c_n(t_0) \, u_n(\mathbf{x}') \exp \left[-i \frac{E_n(t - t_0)}{\hbar} \right], \tag{5.3}$$

where $u_n(\mathbf{x}') \;=\; \langle \mathbf{x}' | a_n \rangle$ (5.4)

is the eigenfunction of the operator $\hat{A}$ with eigenvalues a_n. Also the expansion coefficient $c_n(t_0)$ is

$$c_n(t_0) \;=\; \langle a_n | \alpha, t_0 \rangle \;=\; \int d^3\mathbf{x}' \langle a_n | \mathbf{x}' \rangle \langle \mathbf{x}' | \alpha, t_0 \rangle \tag{5.5}$$

$$=\; \int d^3\mathbf{x}' u_n^*(\mathbf{x}') \, \psi(\mathbf{x}', t_0). \tag{5.6}$$

We thus can write Eq. (5.2) as follows

$$\langle \mathbf{x}'' | \alpha, t_0; t \rangle \;=\; \psi(\mathbf{x}'', t)$$

$$=\; \int d^3\mathbf{x}' \left(\sum_{a_n} \langle \mathbf{x}'' | a_n \rangle \langle a_n | \mathbf{x}' \rangle \exp \left[-i \frac{E_n(t - t_0)}{\hbar} \right] \right) \psi(\mathbf{x}', t_0)$$

$$=\; \int d^3\mathbf{x}' K(\mathbf{x}'', t; \mathbf{x}', t_0) \, \psi(\mathbf{x}', t_0), \tag{5.7}$$

where $K(\mathbf{x}'', t; \mathbf{x}', t_0) \;=\; \sum_{a_n} \langle \mathbf{x}'' | a_n \rangle \langle a_n | \mathbf{x}' \rangle \exp \left[-i \frac{E_n(t - t_0)}{\hbar} \right]$ (5.8)

is the *kernel* of the integral operator which acting on the initial wavefunction yields the final wavefuntion. Thus, the time evolution of the wavefunction can be completely predicted if $K(\mathbf{x}'', t; \mathbf{x}', t_0)$ is known and $\psi(\mathbf{x}', t_0)$ is given initially. In this sense Schrödinger wave mechanics is a perfectly causal theory, provided that the system is left undisturbed. If, however, a measurement intervenes, the wavefunction changes abruptly in an uncontrollable way into one of the eigenfunctions of the observable being measured.

There are two important properties of K.

(i). For $t > t_0$, $K(\mathbf{x}'', t; \mathbf{x}', t_0)$ satisfies Schrödinger's time-dependent wave equation in the variables $\mathbf{x}''$ and t, with $\mathbf{x}'$ and t_0 fixed. This is evident from Eq. (5.8) because $\langle \mathbf{x}'' | a_n \rangle \exp \left[-i \frac{E_n(t - t_0)}{\hbar} \right]$, being the wavefunction corresponding to $\hat{U}(t, t_0) | a_n \rangle$, satisfies the wave equation.

(ii). $\lim_{t \to t_0} K\left(\mathbf{x}'', t; \mathbf{x}', t_0\right) = \delta^3\left(\mathbf{x}'' - \mathbf{x}'\right),$ $\qquad$ (5.9)

which is evident from Eq. (5.8) using $\sum_{a_n} |a_n\rangle\langle a_n| = \hat{I}$.

Because of these two properties, the propagator Eq. (5.8), regarded as a function of $\mathbf{x}''$, is simply the wavefunction at t of a particle which was localized precisely at $\mathbf{x}'$ at an earlier time t_0. This interpretation also follows from Eq. (5.8) if we write

$$K\left(\mathbf{x}'', t; \mathbf{x}', t_0\right) = \left\langle \mathbf{x}'' \left| \exp\left[-i\frac{\hat{H}\left(t - t_0\right)}{\hbar}\right] \right| \mathbf{x}' \right\rangle.$$ $\qquad$ (5.10)

It is evident that the propagator is simply the Green's function for the time dependent wave equation satisfying

$$\left[-\left(\frac{\hbar^2}{2m}\right)\nabla''^2 + V\left(\mathbf{x}''\right) - i\hbar\frac{\partial}{\partial t}\right] K\left(\mathbf{x}'', t; \mathbf{x}', t_0\right) = -i\hbar\delta^3\left(\mathbf{x}'' - \mathbf{x}'\right)\delta\left(t - t_0\right),$$

with the boundary condition $K\left(\mathbf{x}'', t; \mathbf{x}', t_0\right) = 0$ for $t < t_0$.

TRANSITION AMPLITUDE

We have

$$\begin{aligned}
K\left(\mathbf{x}'', t; \mathbf{x}', t_0\right) &= \sum_{a_n}\langle\mathbf{x}''|a_n\rangle\langle a_n|\mathbf{x}'\rangle \exp\left[-i\frac{E_n\left(t - t_0\right)}{\hbar}\right] \\
&= \sum_{a_n}\langle\mathbf{x}''| \exp\left[-i\frac{\hat{H}t}{\hbar}\right] |a_n\rangle\langle a_n| \exp\left[i\frac{\hat{H}t_0}{\hbar}\right] |\mathbf{x}'\rangle \\
&= \langle\mathbf{x}'', t|\mathbf{x}', t_0\rangle,
\end{aligned}$$ $\qquad$ (5.11)

where $|\mathbf{x}', t_0\rangle$ and $\langle\mathbf{x}'', t|$ are to be understood as a eigenket and eigenbra of the position operator in Heisenberg picture. We can then identify $\langle\mathbf{x}'', t|\mathbf{x}', t_0\rangle$ as the probability amplitude for the particle prepared at t_0 with position eigenvalue $\mathbf{x}'$ to be found at a later time t at $\mathbf{x}''$. Or in other words $\langle\mathbf{x}'', t|\mathbf{x}', t_0\rangle$ is the *transition amplitude* from space-time point $(\mathbf{x}', t_0)$ to $(\mathbf{x}'', t)$.

We use symmetric notation and write $\langle\mathbf{x}'', t''|\mathbf{x}', t'\rangle$ as the transition amplitude. We can also use the identity operator

$$\int d^3\mathbf{x}''|\mathbf{x}'', t''\rangle\langle\mathbf{x}'', t''| = \hat{I}.$$

We can then divide the time interval (t', t''') into two parts (t', t'') and (t'', t''') and introduce the identity oerator in between as follows

$$\langle\mathbf{x}''', t'''|\mathbf{x}', t'\rangle = \int d^3\mathbf{x}''\langle\mathbf{x}''', t'''|\mathbf{x}'', t''\rangle\langle\mathbf{x}'', t''|\mathbf{x}', t'\rangle, \quad \text{for} \quad t''' > t'' > t'.$$ (5.12)

This is the composition property of the transition amplitude.

Dividing in still smaller subintervals we can have

$$\langle \mathbf{x}_4, t_4 | \mathbf{x}_1, t_1 \rangle = \int d^3 \mathbf{x}_3 \int d^3 \mathbf{x}_2 \langle \mathbf{x}_4, t_4 | \mathbf{x}_3, t_3 \rangle \langle \mathbf{x}_3, t_3 | \mathbf{x}_2, t_2 \rangle \langle \mathbf{x}_2, t_2 | \mathbf{x}_1, t_1 \rangle$$

$$\text{for} \quad t_4 > t_3 > t_2 > t_1,$$

$$\vdots \quad \vdots \quad \vdots$$

$$\langle \mathbf{x}_N, t_N | \mathbf{x}_1, t_1 \rangle = \int d^3 \mathbf{x}_{N-1} \int d^3 \mathbf{x}_{N-2} \cdots \int d^3 \mathbf{x}_3 \int d^3 \mathbf{x}_2 \langle \mathbf{x}_N, t_N | \mathbf{x}_{N-1}, t_{N-1} \rangle \cdot$$

$$\cdot \langle \mathbf{x}_{N-1}, t_{N-1} | \mathbf{x}_{N-2}, t_{N-2} \rangle \cdots \langle \mathbf{x}_3, t_3 | \mathbf{x}_2, t_2 \rangle \langle \mathbf{x}_2, t_2 \rangle | \mathbf{x}_1, t_1 \rangle,$$

$$\text{for} \quad t_N > t_{N-1} > t_{N-2} > \cdots > t_3 > t_2 > t_1. \quad (5.13)$$

This can be visualized as in Fig. (5.1) by plotting (x_j, t_j) on the space-time plane in one dimension. The initial and the final space-time points are fixed to be (x_1, t_1)

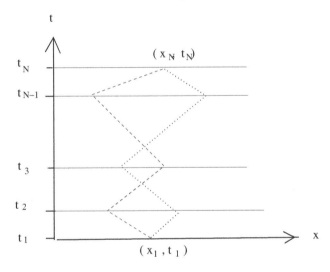

Figure 5.1: Schematic diagram of Feynman's Path in one dimension.

and (x_N, t_N) respectively. Thus to get the transition amplitude between (x_1, t_1) and (x_N, t_N) we have to sum over all possible paths in the space-time plane with the end points fixed.

5.2   Feynman's Path

In classical mechanics a definite path in (x, t) plane is associated with the motion of the particle between two fixed end points which minimizes *Action*. In Feynman formulation of quantum mechanics all possible paths must be included in the integrals, even those which do not bear any resemblance with the classical paths. We, however, have to ensure that the quantum formulation should be able to yield smoothly the classical mechanics in the limit $\hbar \rightarrow 0$.

Introducing the following notation for classical *Action*

$$S(n, n-1) = \int_{t_{n-1}}^{t_n} dt \, L(x, \dot{x}), \tag{5.14}$$

where L is the classical Lagrangian. Since it is a function of x and $\dot{x}$, $S(n, n-1)$ is defined only after a definite path is specified, so that the integration can be carried out. We consider a small segment along the path say between (x_{n-1}, t_{n-1}) and (x_n, t_n). According to Dirac we are to associate $\exp\left[\frac{i}{\hbar} S(n, n-1)\right]$ with this segment. Then going along the definite path we successively multiply expressions of this type to get

$$\Pi_{n=2}^{N} \exp\left[\left(\frac{i}{\hbar}\right) S(n, n-1)\right] = \exp\left[\left(\frac{i}{\hbar}\right) \sum_{n=2}^{N} S(n, n-1)\right]$$

$$= \exp\left[iS(N, 1)/\hbar\right]. \tag{5.15}$$

To get $\langle x_N, t_N | x_1.t_1 \rangle$, we must yet integrate over $x_2 \, x_3, \cdots, x_N$. At the same time, using the composition property, we let the time interval between t_{n-1} and t_n be infinitesimally small. In some loose sense for $\langle x_N, t_N | x_1, t_1 \rangle$ we may write

$$\langle x_N, t_N | x_1, t_1 \rangle \sim \sum_{\text{all paths}} \exp\left[iS(n, n-1)\right]. \tag{5.16}$$

We now check whether the development up to Eq. (5.16) makes any sense in the classical limit $\hbar \rightarrow 0$. As $\hbar \rightarrow 0$, the exponential oscillates very violently, so there is a tendency for cancellation among various contributions from the neighbouring paths. This is because $\exp[iS/\hbar]$ for some definite path and $\exp[iS/\hbar]$ for a slightly different path have large difference in phases because $\hbar \rightarrow 0$. So most of the paths do not contribute in the limit. However for a path along which *Action* is minimum, we have

$$\delta S(N, 1) = 0, \tag{5.17}$$

where the change in S is due to slight deformation of the path with the end points fixed. We call the *Action* $S_{\min}$ which is the classical path. For any other path near about this path with the end points same, *Action* is very nearly equal to $S_{\min}$. As a result, near the classical path constructive interference between neighbouring paths is possible as the phases do not change much.

To formulate Feynman's conjecture more precisely we go back to $\langle x_n, t_n | x_{n-1}, t_{n-1}\rangle$, where $\Delta t = t_n - t_{n-1}$ is assumed to be infinitesimally small and write

$$\langle x_n, t_n | x_{n-1}, t_{n-1}\rangle = \left[\frac{1}{\omega(\Delta t)}\right] \exp\left[\frac{iS(n.n-1)}{\hbar}\right]. \tag{5.18}$$

We have to evaluate $S(n, n-1)$ in the limit $\Delta t \to 0$. The weight factor $1/\omega(\Delta t)$ which is assumed to depend only on the time interval Δt and not on $V(x)$. This factor is needed from dimensional considertion, according to the way we normalized our position eigenkets. $\langle x_n, t_n | x_{n-1}, t_{n-1}\rangle$ must have the dimension of inverse of length.

Now we make a straight line approximation to the path joinng (x_{n-1}, t_{n-1}) and $(x_n.t_n)$ as follows (since Δt is small)

$$S(n, n-1) = \int_{t_{n-1}}^{t_n} dt \left[\frac{m\dot{x}^2}{2} - V(x)\right]$$

$$= \Delta t \left\{\frac{m}{2}\left(\frac{x_n - x_{n-1}}{\Delta t}\right)^2 - V\left(\frac{x_n + x_{n-1}}{2}\right)\right\}. \tag{5.19}$$

For free particles $V = 0$ and Eq. (5.18) then becomes

$$\langle x_n, t_n | x_{n-1}, t_{n-1}\rangle = \left[\frac{1}{\omega(\Delta t)}\right] \exp\left[\frac{im(x_n - x_{n-1})^2}{2\hbar\Delta t}\right] \tag{5.20}$$

Problem 5.1 Use the following free particle wavefunction

$$\langle x''|p'\rangle = \frac{1}{\sqrt{2\pi\hbar}} \exp\left[\frac{ip'x''}{\hbar}\right], \quad \text{and} \quad \hat{H}|p'\rangle = \frac{p'^2}{2m}|p'\rangle,$$

calculate the free particle propagator from Eq. (5.8) and show that

$$K(x'', t; x', t_0) = \sqrt{\frac{m}{2\pi i\hbar(t - t_0)}} \exp\left[\frac{im(x'' - x')^2}{2\hbar(t - t_0)}\right]. \tag{5.21}$$

Thus the exponent in Eq. (5.20) is the same as that of the free particle propagator.

Since $\frac{1}{\omega(\Delta t)}$ is independent of $V(x)$ it can be worked out from free particle propagator and is given by

$$\frac{1}{\omega(\Delta t)} = \sqrt{\frac{m}{2\pi i\hbar\Delta t}}.$$

Thus as $\Delta t \to 0$ we have

$$\langle x_n, t_n | x_{n-1}, t_{n-1} \rangle = \sqrt{\frac{m}{2\pi i \hbar \Delta t}} \exp\left[\frac{iS(n, n-1)}{\hbar}\right]. \tag{5.22}$$

And finally with $(t_N - t_1)$ finite

$$\langle x_N, t_N | x_1, t_1 \rangle = \lim_{N \to \infty} \left(\frac{m}{2\pi i \hbar \Delta t}\right)^{(N-1)/2} \int dx_{N-1} \int dx_{N-2} \cdots \int dx_3 \int dx_2$$

$$\Pi_{n=2}^N \exp\left[\frac{iS(n, n-1)}{\hbar}\right], \tag{5.23}$$

and the limit $N \to \infty$ is taken with x_N, t_N fixed.

We can define a new kind of multidimensional (infinite dimensional) integral operator as follows

$$\int_{x_1}^{x_N} \mathcal{D}[x(t)] \equiv \lim_{N \to \infty} \left(\frac{m}{2\pi i \hbar \Delta t}\right)^{(N-1)/2} \int dx_{N-1} \int dx_{N-2} \cdots \int dx_3 \int dx_2,$$

$$\text{together with } \Delta t \to 0 \text{ and } N \cdot \Delta t = t_N - t_1, \tag{5.24}$$

and write Eq. (5.23) as

$$\langle x_N, t_N | x_1, t_1 \rangle = \int_{x_1}^{x_N} \mathcal{D}[x(t)] \exp\left[i \int_{t_1}^{t_N} dt \frac{L_{\rm cl}(x, \dot{x})}{\hbar}\right]. \tag{5.25}$$

This expression is known as Feynman's *path integral*, where the sum over all possible paths is evident.

Chapter 6

Application in One Dimension

Study of one dimensional problems is of interest not only because several physical situations are effectively one dimensional but also because a number of more complicated problems can be reduced to the solutions of equations similar to one dimensional Schrödinger equation.

The *time dependent Schrödinger equation* for a particle of mass m moving in a potential $V(x)$ is given by

$$i\hbar\frac{\partial}{\partial t}\Psi(x,t) = \left[-\frac{\hbar^2}{2m}\frac{\partial^2}{\partial x^2} + V(x)\right]\Psi(x,t). \tag{6.1}$$

If V is time independent we can look for a *stationary state* solution of Eq. (6.1) in the form

$$\Psi(x,t) = \psi(x)\exp\left[-\frac{iEt}{\hbar}\right], \tag{6.2}$$

where E is the energy for the stationary state. The *time-independent Schrödinger equation* is

$$-\frac{\hbar^2}{2m}\frac{d^2\psi(x)}{dx^2} + V(x)\psi(x) = E\psi(x). \tag{6.3}$$

6.1   Free Particle

The time-independent Schrödinger equation is

$$-\frac{\hbar^2}{2m}\frac{d^2\psi(x)}{dx^2} + V(x)\psi(x) = E\psi(x), \tag{6.4}$$

$$\frac{d^2\psi(x)}{dx^2} + k^2\psi(x) = 0, \tag{6.5}$$

$$\text{where} \qquad\qquad k^2 = \frac{2mE}{\hbar^2}. \tag{6.6}$$

The general solution is

$$\psi\left(x\right) \;=\; A\exp\left(ikx\right) + B\exp\left(-ikx\right).\tag{6.7}$$

For a physically acceptable solution k cannot have any imaginary part because in that case $\psi\left(x\right)$ will blow up either at $x = \infty$ or at $x = -\infty$. In other words

$$E \;=\; \frac{k^2\hbar^2}{2m} \;\geq\; 0,\tag{6.8}$$

and because any non-negative value of E is allowed, the energy spectrum is continuous extending from $E = 0\ E = \infty$. This is also obvious since for a free article E is the kinetic energy.

The general solution of the Schrödinger equation in a stationary state for $E > 0$ is given by

$$\begin{aligned}\Psi\left(x,t\right) &\;=\; \left(Ae^{ikx} + Be^{-ikx}\right)\exp\left[-iEt/\hbar\right]\\ &\;=\; Ae^{i(kx-\omega t)} + Be^{-i(kx+\omega t)},\end{aligned}\tag{6.9}$$

where $\omega = E/\hbar$ is the angular frequency. Considering the particular case $B = 0$, the plane wave

$$\Psi\left(x.t\right) \;=\; Ae^{i(kx-\omega t)}\tag{6.10}$$

is the momentum eigenfunction representing an oscillatory travelling wave in a positive $x-$direction with a definite momentum $p = k\hbar$ and a *phase velocity* $v_{\mathrm{ph}} = \frac{\omega}{k} = \frac{k\hbar}{2m}$. However, the particle velocity $v = \frac{p}{m} = \frac{k\hbar}{m}$ is *not* equal to the phase velocity, but is equal to the *group valocity* $v_{\mathrm{g}} = \frac{d\omega}{dk}$ of the plane wave. The *angular frequency* is $\omega = E/\hbar$ and the *wave number* $k = \frac{p}{\hbar} = \frac{2\pi}{\lambda}$, λ being the *de Broglie wavelength* of the particle. The *probability density*

$$P \;=\; |\Psi\left(x,t\right)|^2 \;=\; |A|^2\tag{6.11}$$

which is independent of time (as for any stationary state solution) and is also independent of x, so that the position of the particle on the x-axis is completely unknown.

This is in accordance with the *Heisenberg Uncertainty Principle*, since the particle has a definite momentum its position cannot be localized on the x-axis.

BOX NORMALIZATION

Since the integral $\int_{-\infty}^{+\infty}\Psi^*\left(x,t\right)\Psi\left(x,t\right)\ dx$ for the Ψ of Eq. (6.10) will become infinity we restrict the domain of $\psi(x)$ to an arbitrary large one dimensional 'box' of length L such that $\psi(x)$ satisfies periodic boundary considerations at the walls,

$$\psi(x+L) = \psi(x). \tag{6.12}$$

As a consequence k gets *quantized*

$$k_n = \frac{2\pi}{L}n, \quad \text{with} \quad n = 0, \pm 1, \pm 2, \pm 3, \cdots. \tag{6.13}$$

The state corresponding to $n = 0$ is the *ground state*. The energy spectrum becomes discrete

$$E_n = \frac{k^2 \hbar^2}{2m} = \frac{2\pi^2 \hbar^2}{mL^2}n^2, \tag{6.14}$$

each eigenvalue (except $E = 0$ corresponding to $n = 0$) being doubly degenerate.

The normalized eigenfunctions for the free particle is thus

$$\psi_k(x) = \frac{1}{\sqrt{L}} \exp[ikx] \tag{6.15}$$

which are also orthogonal, because we have

$$\int_{-L/2}^{+L/2} \psi_{k'}^*(x)\,\psi_k(x)\,dx = \frac{1}{L}\int_{-L/2}^{+L/2} \exp[i(k-k')x]\,dx = \delta_{k,k'}, \tag{6.16}$$

where we have used Eq. (6.13).

DELTA FUNCTION NORMALIZATION

We can also use the definition of Delta Function[1] to set up a delta function normalization.

Thus using Eq. (A.28) we have

$$\int_{-\infty}^{+\infty} \exp[-i(k-k')x]\,dx = 2\pi\delta(k-k'). \tag{6.17}$$

Normalized eigenfunctions are

$$\psi_k(x) = \frac{1}{\sqrt{2\pi}} \exp[ikx]. \tag{6.18}$$

The *closure property* is then given by

$$\int_{-\infty}^{+\infty} \psi_k^*(x')\,\psi_k(x)\,dk = \delta(x-x'). \tag{6.19}$$

---

[1]See Chapter A.1,

WAVE PACKETS

We have seen that the free particle plane wave is not localized. We can however construct a wave packet by superposing plane waves given in Eq. (6.18) as follows

$$\Psi\left(x,t\right) \;=\; \frac{1}{\sqrt{2\pi}}\int_{-\infty}^{+\infty} A\left(k\right)e^{+i\left[kx-\omega\left(k\right)t\right]}\,dk, \tag{6.20}$$

$$\text{where}\quad \omega\left(k\right) \;=\; \frac{\hbar k^2}{2m}.$$

We take the amplitude function $A\left(k\right)$ sharply peaked near $k = k_0$ and integrate Eq. (6.20) about k_0:

$$\Psi\left(x,t\right) \;=\; \frac{1}{\sqrt{2\pi}}\int_{k_0-\Delta k}^{k_0+\Delta k} A\left(k\right)\exp\left[i\left(kx-\omega\left(k\right)t\right)\right]\,dk \tag{6.21}$$

Introducing the variable $\xi \;=\; k - k_0$

$$\omega\left(k\right) \;\approx\; \omega\left(k_0+\xi\right) \;=\; \omega\left(k_0\right) \;+\; \left(\frac{d\omega}{dk}\right)_0 \xi \;=\; \omega_0 \;+\; \left(\frac{d\omega}{dk}\right)_0 \xi,$$

we get after integration

$$\Psi\left(x,t\right) \;=\; A\left(x,t\right)\exp\left[i\left(k_0 x - \omega_0 t\right)\right]. \tag{6.22}$$

The amplitude function $A\left(x,t\right)$ of the plane wave $\exp\left[i\left(k_0 x - \omega_0 t\right)\right]$ is given by

$$A\left(x,t\right) \;=\; \frac{2A\left(k_0\right)\sin\left\{\left[x-\left(\frac{d\omega}{dk}\right)_0 t\right]\Delta k\right\}}{\left[x-\left(\frac{d\omega}{dk}\right)_0 t\right]}, \tag{6.23}$$

whose maximum at $t = 0$ occurs at the origin $x = 0$ and is $2A\left(k_0\right)\Delta k$. Thus at $t = 0$ we get

$$A\left(x,t=0\right) \;=\; A_0 \frac{\sin\theta}{\theta} \;=\; A_0 f\left(\theta\right),$$

$$\text{where}\quad A_0 \;=\; 2A\left(k_0\right)\Delta k,$$

$$\text{and}\quad \theta \;=\; x\cdot\Delta k.$$

The amplitude function $A\left(x,t=0\right)$ is zero when

$$x_n \;=\; \frac{\theta_n}{\pi} \;=\; \frac{n\pi}{\Delta k}, \quad n \;=\; \pm 1,\ \pm 2,\ \cdots.$$

The *spatial extension of the packet* is

$$\Delta x \;=\; \left(x_{+1}\right) - \left(x_{-1}\right) \;=\; 2x_1 \;=\; \frac{2\pi}{\Delta k}. \tag{6.24}$$

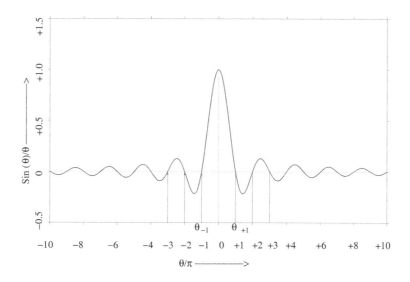

Figure 6.1: Plot of $f(\theta)$ as a function of θ.

Since

$$\Delta k \;=\; \frac{\Delta p}{\hbar}, \text{ so } \Delta x \cdot \Delta p \;=\; 2\pi\hbar. \tag{6.25}$$

With passage of time the mid-point of the wave packet moves with *group velocity*

$$v_g \;=\; \left(\frac{d\omega}{dk}\right)_{k_0} \;=\; \frac{p_0}{m}.$$

which is also the particle velocity.

6.2   Rectangular Potential Well

This potential is dfined as

$$V(x) \;=\; \begin{cases} 0, & \text{for } -a/2 \;<\; x \;<\; +a/2 \\ V_0, & \text{elsewhere} \end{cases}. \tag{6.26}$$

Rectangular Potential Well

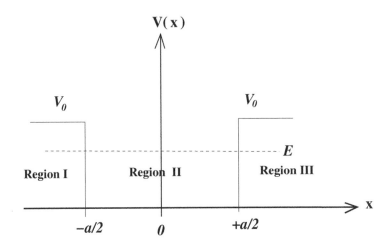

Figure 6.2: The rectangular potential well in 1 dimension.

(i)   Bound State Solution

For bound state $0 < E < V_0$ the Schrödinger equation is given by

$$\frac{d^2\psi_{II}}{dx^2} + k^2\psi_{II} = 0, \quad \text{for } |x| < a/2,$$

$$\frac{d^2\psi_{I,III}}{dx^2} - \gamma^2\psi_{I,III} = 0, \quad \text{for } |x| > a/2, \tag{6.27}$$

$$\text{where } k^2 = \frac{2mE}{\hbar^2}, \quad \text{and } \gamma^2 = \frac{2m(V_0 - E)}{\hbar^2}. \tag{6.28}$$

Here $\psi_{I,III}$ refer to solutions in the Regions I and III, and ψ_{II} refers to solution in Region II shown in Fig. (6.2).

The potential and the kinetic energies are invariant under reflection $x \to -x$ and thus the Hamiltonian is invariant under *Parity operation*[2]. So both $\psi(x)$ and $\psi(-x)$ are solutions of the Schrödinger equation. We can also show that in the case of one dimensional bound states (*i.e* discrete energy spectrum) none of the energy levels are degenerate.

Proof:

If possible let ψ_1 and ψ_2 are two solutions corresponding to the same energy E.

---

[2]See Chapter 11.

Then

$$\psi_1'' = \frac{2m}{\hbar^2}(V - E)\psi_1$$

$$\psi_2'' = \frac{2m}{\hbar^2}(V - E)\psi_2$$

Here 'prime' refers to differentiation with respect to x. Thus

$$\frac{\psi_1''}{\psi_1} = \frac{\psi_2''}{\psi_2}$$

$$\text{or } \psi_2\psi_1'' - \psi_1''\psi_2 = 0,$$

$$\psi_2\psi_1' - \psi_1\psi_2' = \text{Constant.}$$

Since for bound states both $\psi_1, \psi_2 \to 0$ at $x \to \pm\infty$, so the integration constant is zero and we have

$$\frac{\psi_1'}{\psi_1} = \frac{\psi_2'}{\psi_2}, \quad \text{or} \quad \psi_1 = \text{Constant} \times \psi_2,$$

thus ψ_1 and ψ_2 turns out to be doubly-degenerate.

Now since the potential in this problem is an even function of x, the solutions of the Schrödinger's equation have definite parity, hence need be determined only for positive values of x.

EVEN PARITY SOLUTIONS

Even parity solutions are given by

$$\psi(x) = B\cos(kx), \quad 0 < x < a/2, \tag{6.29}$$

$$\psi(x) = Ae^{-\gamma x}, \quad x \ge a/2.$$

Continuity of ψ and $\frac{d\psi}{dx}$ at $x \ge a/2$ yields

$$B\cos\left(\frac{ka}{2}\right) = Ae^{-\gamma a/2}, \tag{6.30}$$

$$-kB\sin\left(\frac{ka}{2}\right) = -A\gamma e^{-\gamma a/2}$$

$$\text{or} \quad \tan\left(\frac{ka}{2}\right) = \frac{\gamma}{k} = \sqrt{\frac{2mV_0}{\hbar^2 k^2} - 1}. \tag{6.31}$$

Now let p be largest integer in $\frac{ka}{2\pi}$, then

$$p\pi \le \frac{ka}{2} < (p+1)\pi, \quad \text{i.e.} \quad 0 \le \left(\frac{ka}{2} - p\pi\right) < \pi \tag{6.32}$$

and we also have

$$\tan\left(\frac{ka}{2}\right) = \tan\left(\frac{ka}{2} - p\pi\right).$$

Now if $\pi/2 < \left(\frac{ka}{2} - p\pi\right) < \pi$, then $\tan\left(\frac{1}{2}ka\right)$ is negative and Eq. (6.31) cannot hold. Thus we must have

$$0 \le \left(\frac{ka}{2} - p\pi\right) < \pi/2, \quad \text{or} \quad p\pi \le \frac{1}{2}ka < \left(p + \frac{1}{2}\right)\pi. \tag{6.33}$$

The results of Eq. (6.31) and Eq. (6.33) is geometrically depicted in Fig. (6.3). From Fig. (6.3) we have

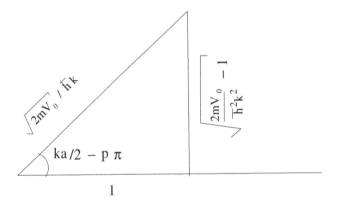

Figure 6.3: Geometrical depiction of Eq. (6.31) and Eq. (6.33).

$$
\begin{aligned}
\cos\left(\frac{ka}{2} - p\pi\right) &= \frac{k\hbar}{\sqrt{2mV_0}} \\
&= \sin\left[\pi/2 - \left(\frac{ka}{2} - p\pi\right)\right] \\
&= \sin\left[\left(p + \frac{1}{2}\right)\pi - \frac{ka}{2}\right].
\end{aligned}
$$

$$\text{So} \quad \left(p + \frac{1}{2}\right)\pi - \frac{ka}{2} = \arcsin\frac{k\hbar}{\sqrt{2mV_0}},$$

$$\text{or} \quad ka = (2p+1)\pi - 2\arcsin\frac{k\hbar}{\sqrt{2mV_0}}, \tag{6.34}$$

$$\text{where} \quad p = 1, 2, \cdots.$$

Now from Eq. (6.33) the maximum value of $\frac{ka}{2}$ is $\left(p + \frac{1}{2}\right)\pi$ and the minimum value of $\frac{ka}{2}$ is $p\pi$. The minimum value of arcsin $\frac{k\hbar}{\sqrt{2mV_0}}$ is 0 and the maximum value is $\pi/2$, i.e.

$$0 \leq \text{arcsin} \frac{k\hbar}{\sqrt{2mV_0}} \leq \pi/2. \tag{6.35}$$

Since Eq. (6.34) will have *no* solution unless the argument of arcsin is ≤ 1 i.e.

$$k^2\hbar^2 \leq 2mV_0, \quad \text{and} \quad E = \frac{k^2\hbar^2}{2m} < V_0.$$

Thus we have for *even* solutions $k_p^{(+)}$, for $p = 0, 1, 2, \cdots, , p_{\text{max}}^+$, corresponding to the points of intersection of the straight line ka and the *monotonically drcreasing* curves

$$\zeta_{2p+1}(k) = (2p+1)\pi - 2\,\text{arcsin}\frac{k\hbar}{\sqrt{2mV_0}}. \tag{6.36}$$

Also from Eq. (6.34) putting the maximum and minimum values of arcsin $\frac{k\hbar}{\sqrt{2mV_0}}$,

$$2p\pi \leq k_p^+ a \leq (2p+1)\pi;$$

the + sign refers to even parity solution. Now $k_p^+ < \frac{\sqrt{2mV_0}}{\hbar}$ and for $\sqrt{2mV_0} \gg k\hbar$ Eq. (6.34) takes the following form (since for small values of x, arcsin$(x) \approx x$):

$$
\begin{aligned}
k^+ a &= (2p+1)\pi, \quad \text{and} \\
k_{p\text{max}}^+ &= \frac{\pi}{a}(2p+1) \\
E^+ &= \frac{k_{\text{max}}^2 \hbar^2}{2m} = \frac{\pi^2\hbar^2}{2ma^2}(2p+1)^2 \\
\text{with} \quad p &= 0,\ 1,\ 2,\ \cdots,
\end{aligned}
\tag{6.37}
$$

Also $\psi(x) \to 0$ in this limit in the region $x > a/2$.

The normalized wavefunction is given by

$$
\begin{aligned}
\psi^+(x) &= \sqrt{\frac{2}{a}}\cos\left(\frac{\pi}{a}(2p+1)x\right), \\
p &= 0,\ 1,\ 2,\ \cdots
\end{aligned}
\tag{6.38}
$$

$$\text{and} \quad 0 < \text{arcsin}\frac{k\hbar}{\sqrt{2mV_0}} \leq \pi/2.$$

ODD PARITY SOLUTIONS

$$\psi^-(x) = \begin{cases} C\sin\left(\frac{1}{2}kx\right), & \text{for } x < a/2 \\ A\exp\left(-\gamma x\right), & \text{for } x > a/2 \end{cases}.$$

Then continuity of ψ and $\frac{d\psi}{dx}$ at $x = a/2$ leads finally to the following:

$$\cot\left(\frac{ka}{2}\right) = -\frac{\gamma}{k} = -\sqrt{\frac{2mV_0}{k^2\hbar^2} - 1}. \tag{6.39}$$

Proceeding in similar way as was done in the even parity case, and remembering

$$\cot\left(\frac{ka}{2}\right) = \cot\left(\frac{ka}{2} - p\pi\right)$$

we have if $0 \le \left(\frac{ka}{2} - p\pi\right) < \pi/2$, then $\cot\left(\frac{1}{2}ka\right) \ge 0$, so that Eq. (6.39) cannot hold. Thus we must have

$$\left(p + \frac{1}{2}\right)\pi \le \frac{ka}{2} < (p+1)\pi. \tag{6.40}$$

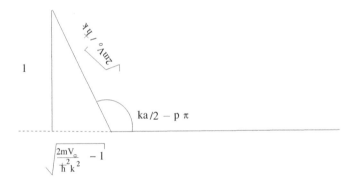

Figure 6.4: Geometrical depiction of Eq. (6.39) and Eq. (6.40).

From Fig. (6.4), we get

$$\sin\left[(p+1)\pi - \frac{ka}{2}\right] = \frac{k\hbar}{\sqrt{2mV_0}},$$

$$(p+1)\pi - \frac{ka}{2} = \arcsin\frac{k\hbar}{\sqrt{2mV_0}},$$

$$ka = (2p+2) - 2\arcsin\frac{k\hbar}{\sqrt{2mV_0}} \tag{6.41}$$

$$\text{with } p = 0,\ 1,\ 2,\ \cdots$$

$$\text{and } 0 < \arcsin\frac{k\hbar}{\sqrt{2mV_0}} \le \pi/2.$$

We have *odd parity solutions*

$k_p^{(-)}$ with $p = 0,\ 1,\ 2,\ \cdots,\ p_{\max}^-,$

corresponding to the points of intersection of the straight line ka and the *monotonically decreasing curves*

$$\zeta_{2p+2}(k) \;=\; (2p+2)\,\pi - 2\arcsin\frac{k\hbar}{\sqrt{2mV_0}}. \tag{6.42}$$

We see that

$$(2p+1)\,\pi \le k_p^{(-)}a < (2p+2)\,\pi \quad\text{and}\quad k_p^- < \frac{\sqrt{2mV_0}}{\hbar}.$$

We can now combine Eq. (6.34) and Eq. (6.41) and write

$$ka \;=\; n\pi - 2\arcsin\frac{k\hbar}{\sqrt{2mV_0}}, \tag{6.43}$$

$$\text{where}\quad n \;=\; \begin{cases} 1,\ 3,\ 5,\ \cdots,\ (2p+1) & \text{for even parity solutions} \\ 2,\ 4,\ 6,\ \cdots,\ (2p+2) & \text{for odd parity solutions} \end{cases}.$$

Graphical solution of Eq. (6.43) is shown here in Fig. (6.5).

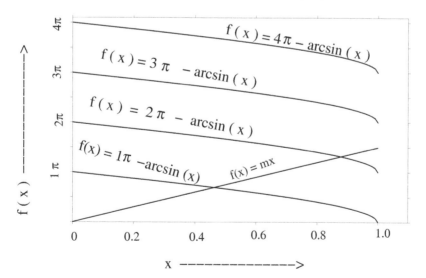

Figure 6.5:   Graphical solution of Eq. (6.43). $f(x) = n\pi - arcsin(x)$ and the straight line is $f(x) = ka = mx$ where $x = \frac{k\hbar}{\sqrt{2mV_0}}$.

(ii)   Unbound State Solution

In this case $0 < V_0 < E$, so the particle is not bound. Assuming that the particle is incident upon the well from the left. the solution of the Schrödinger equation in the external regions $x < -a/2$ and $x > a/2$ are given by

$$\psi(x) = \begin{cases} Ae^{+ikx} + Be^{-ikx} & \text{for} \quad x < -a/2 \\ Ce^{+ikx} & \text{for} \quad x > +a/2 \end{cases} , \tag{6.44}$$

where

$$k = +\frac{\sqrt{2m(E - V_0)}}{\hbar}, \quad \text{and} \quad A, B, C \quad \text{are constants.}$$

Since there is no reflector at large positive values of x there is no *reflected* term of the form e^{-ikx} in the region $x > +a/2$. In the region $x < -a/2$, however, the wavefunction consists of the *incident* wave e^{+ikx} of amplitude A and the *reflected* wave e^{-ikx} of amplitude B, whereas in the region $x > a/2$, it is only the *transmitted* wave of amplitude C. Inside the region $-a/2 < x < +a/2$ the solution is given by

$$\psi(x) = Fe^{+i\alpha x} + Ge^{-i\alpha x} \tag{6.45}$$

$$\text{where } \alpha = +\frac{\sqrt{2mE}}{\hbar}. \tag{6.46}$$

Problem 6.1 Requiring that ψ and $\frac{d\psi}{dx}$ are continuous at $x = \pm a/2$ find the expressions for

(i). Reflection Coefficient

$$R = |B/A|^2, \tag{6.47}$$

(ii). Transmission Coefficient

$$T = |C/A|^2. \tag{6.48}$$

Also check that $T + R = 1$.

6.3   Rectangular Potential Barrier

We shall discuss this one-dimensional case in some details and introduce in an elementary way the *S-matrix* theory of collision.

Rectangular Potential Barrier

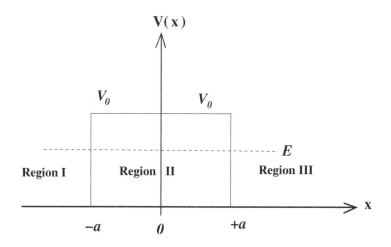

Figure 6.6: The rectangular potential barrier in 1 dimension.

The potential is defined as

$$V(x) = \begin{cases} 0 & \text{for} \quad x < -a & \text{Region I} \\ V_0 & \text{for} \quad -a < x < +a & \text{Region II} \\ 0 & \text{for} \quad +a < x & \text{Region III} \end{cases} . \tag{6.49}$$

Since the quantum mechanical barrier penetration occurs when E is less than V_0, we shall study this particularly important case here.

The particle is free for $x < -a$ and $+a < x$. For this reason, the rectangular barrier simulates schematically, the scattering of a free particle from any potential.

The general solution of the Schrödinger equation for $E < V_0$ is

$$\psi(x) = \begin{cases} Ae^{+ikx} + Be^{-ikx} & \text{for} \quad x < -a \\ Ce^{+\kappa x} + De^{-\kappa x} & \text{for} \quad -a < x < +a \\ Fe^{+ikx} + Ge^{-ikx} & \text{for} \quad +a < x \end{cases} . \tag{6.50}$$

where

where $k\hbar = \sqrt{2mE}$, and $\kappa\hbar = \sqrt{2m(V_0 - E)}$.

From the boundary conditions of continuity of ψ and $\frac{d\psi}{dx}$ at $x = -a$ we have

$$Ae^{-ika} + Be^{+ika} = Ce^{+\kappa a} + De^{-\kappa a} \quad \text{and}$$
$$Ae^{-ika} - Be^{+ika} = \frac{i\kappa}{k}\left(Ce^{+\kappa a} - De^{-\kappa a}\right). \tag{6.51}$$

These linear homogeneous equations can also be expressed in terms of matrices.

$$\begin{pmatrix} A \\ B \end{pmatrix} = \frac{1}{2} \begin{pmatrix} \left(1 + \frac{i\kappa}{k}\right)e^{+\kappa a + ika} & \left(1 - \frac{i\kappa}{k}\right)e^{-\kappa a + ika} \\ \left(1 - \frac{i\kappa}{k}\right)e^{+\kappa a - ika} & \left(1 + \frac{i\kappa}{k}\right)e^{-\kappa a - ika} \end{pmatrix} \begin{pmatrix} C \\ D \end{pmatrix}.$$

Similarly the boundary conditions at $x = a$ yield

$$\begin{pmatrix} C \\ D \end{pmatrix} = \frac{1}{2} \begin{pmatrix} \left(1 - \frac{ik}{\kappa}\right)e^{+\kappa a + ika} & \left(1 + \frac{ik}{\kappa}\right)e^{-\kappa a + ika} \\ \left(1 + \frac{ik}{\kappa}\right)e^{+\kappa a - ika} & \left(1 - \frac{ik}{\kappa}\right)e^{-\kappa a - ika} \end{pmatrix} \begin{pmatrix} F \\ G \end{pmatrix}.$$

Combining these two matrix equations we get

$$\begin{pmatrix} A \\ B \end{pmatrix} = \begin{pmatrix} \left(\cosh 2\kappa a + \frac{i\epsilon}{2}\sinh 2\kappa a\right)e^{+2ika} & +i\frac{\eta}{2}\sinh 2\kappa a \\ -\frac{i\eta}{2}\sinh 2\kappa a & \left(\cosh 2\kappa a - \frac{i\epsilon}{2}\sinh 2\kappa a\right)e^{-2ika} \end{pmatrix}$$
$$\times \begin{pmatrix} F \\ G \end{pmatrix}. \qquad (6.52)$$

where

$$\epsilon = \frac{\kappa}{k} - \frac{k}{\kappa}, \quad \text{and} \quad \eta = \frac{\kappa}{k} + \frac{k}{\kappa}. \qquad (6.53)$$

Problem 6.2 Calculate the determinant of the matrix in Eq. (6.52) and show that it is equal to 1.

Letting $G = 0$ in Eq. (6.52) we obtain the particular solution which represents a wave incident from the left and transmitted through the barrier to the right. A reflected wave whose amplitude is B, is also present in the region $x < -a$.

Problem 6.3 Show that

$$\frac{F}{A} = \frac{e^{-2ika}}{\cosh 2\kappa a + \frac{i\epsilon}{2}\sinh 2\kappa a}. \qquad (6.54)$$

$|F/A|^2$ is the *transmission coefficient* for the barrier. For a high and wide barrier we have $ka \gg 1$ and

$$\cosh 2ka = \sinh 2ka \approx \frac{1}{2}e^{2ka}$$

$$\text{hence} \quad T = \left|\frac{F}{A}\right|^2 \approx 16e^{-4\kappa a}\left(\frac{k\kappa}{k^2 + \kappa^2}\right)^2. \qquad (6.55)$$

The matrix which relates A and B with F and G in Eq. (6.52) has many simple properties. We write the linear relations as

$$\begin{pmatrix} A \\ B \end{pmatrix} = \begin{pmatrix} \alpha_1 + i\beta_1 & \alpha_2 + i\beta_2 \\ \alpha_3 + i\beta_3 & \alpha_4 + i\beta_4 \end{pmatrix} \begin{pmatrix} F \\ G \end{pmatrix} \qquad (6.56)$$

and compare this with Eq. (6.52). We observe that that the eight real numbers α, β in the matrix satisfy the conditions

$$\alpha_1 = \alpha_4, \quad \beta_1 = -\beta_4, \quad \alpha_2 = \alpha_3 = 0, \quad \beta_2 = -\beta_3. \tag{6.57}$$

Thus the matrix in Eq. (6.56) can be written as

$$\begin{pmatrix} \alpha_1 + i\beta_1 & i\beta_2 \\ -i\beta_2 & \alpha_1 - i\beta_1 \end{pmatrix}. \tag{6.58}$$

Using the result of Problem (6.2) we have

$$\alpha_1^2 + \beta_1^2 - \beta_2^2 = 1. \tag{6.59}$$

Hence we have only 2 parameters that define the matrix and they are ka and κa.

It can be shown that the conditions Eq. (6.57) and Eq. (6.58) imposed on Eq. (6.56) are consequences of the general symmetry properties of the physical system being studied.

SYMMETRIES AND INVARIANCE PROPERTIES[3].

Since the rectangular barrier, Fig. (6.6) is a real potential which is symmetrical about the origin, the Schrödinger equation is invariant under time reversal and space reflection. We can use these properties to derive the general form of the matrix linking the incident wave with the transmitted wave.

We once again write the form of the general solution for ready reference

$$\psi(x) = \begin{cases} Ae^{+ikx} + Be^{-ikx} & \text{for } x < -a \\ Ce^{+\kappa x} + De^{-\kappa x} & \text{for } -a < x < +a \\ Fe^{+ikx} + Ge^{-ikx} & \text{for } +a < x \end{cases}.$$

Instead of using the boundary conditions at $x = -a$ and $x = +a$, we regard that the wavefunction on one side of the barrier, say for $x > a$, as given, then this will give linear equation expressing the coefficienta A and B in terms of F and G. Hence we can express these relations in terms of the matrix M such that

$$\begin{pmatrix} A \\ B \end{pmatrix} = \begin{pmatrix} M_{11} & M_{12} \\ M_{21} & M_{22} \end{pmatrix} \begin{pmatrix} F \\ G \end{pmatrix}. \tag{6.60}$$

Equivalently we can express coefficients of the outgoing waves B and F in terms of the coefficients A and G of the incoming waves which are then known

$$\begin{pmatrix} B \\ F \end{pmatrix} = \begin{pmatrix} S_{11} & S_{12} \\ S_{21} & S_{22} \end{pmatrix} \begin{pmatrix} A \\ G \end{pmatrix}, \tag{6.61}$$

---

[3]See Chapter 11

where S is known as the *scattering matrix* or the *S-matrix* .

The Eq. (6.61) describes the scattering phenomenon when knowing the incoming waves one can calculate the outgoing waves. The symmetry properties can be well formulated in terms of the S-matrix.

The S and the M matrices can be simply related if conservation of probability is invoked.

In an one dimensional stationary state, the probability current density must be independent of x.

$$j = \frac{\hbar}{2mi} \left[\psi^* \frac{d\psi}{dx} - \frac{d\psi^*}{dx} \psi \right]$$

for the stationary state[4] $\frac{dj}{dx} = 0$ and j has the same value at all point x. This gives

$$|A|^2 - |B|^2 = |F|^2 - |G|^2,$$
$$\text{or } |B|^2 + |F|^2 = |A|^2 + |G|^2.$$

This is expected since $|A|^2$ and $|F|^2$ measure the probability current to the right, while $|B|^2$ and $|G|^2$ measure the flow in the opposite direction. In matrix notation we can write

$$(B^* \; F^*) \begin{pmatrix} B \\ F \end{pmatrix} = (A^* \; G^*) \tilde{S}^* S \begin{pmatrix} A \\ G \end{pmatrix}$$
$$= (A^* \; G^*) \begin{pmatrix} A \\ G \end{pmatrix}$$

$$\text{or } \tilde{S}^* S = \hat{I}, \quad \text{a unit matrix.}$$

Thus S is unitary.

Problem 6.4 Use the condition $S^\dagger S = S S^\dagger = \hat{I}$ to prove the following.

$$|S_{11}| = |S_{22}| \quad \text{and} \quad |S_{12}| = |S_{21}|, \tag{6.62}$$
$$|S_{11}|^2 + |S_{12}|^2 = 1, \tag{6.63}$$
$$S_{11} S_{12}^* + S_{21} S_{22}^* = 0. \tag{6.64}$$

Since unitary matrices are extremely important in quantum mechanical formalism, we start with an S of the form

$$S = \begin{pmatrix} u e^{i\alpha} & v e^{i\beta} \\ x e^{i\delta} & w e^{i\gamma} \end{pmatrix}$$

with

$$0 \leq u, v, w, x, \quad \text{and} \quad -\pi \leq \alpha, \beta, \gamma, \delta < +\pi.$$

---

[4]from continuity condition, putting $\frac{\partial \rho}{\partial t} = 0$..

Problem 6.5 Show that S will have only one of the following forms.

I. $$S = \begin{pmatrix} e^{i\alpha} & 0 \\ 0 & e^{i\gamma} \end{pmatrix}, \quad \text{or} \quad S = \begin{pmatrix} 0 & e^{i\beta} \\ e^{i\delta} & 0 \end{pmatrix}.$$

II. $$S = \begin{pmatrix} ue^{i\alpha} & \sqrt{1-u^2}e^{i\beta} \\ -\sqrt{1-u^2}e^{i(\alpha+\gamma+\beta)} & ue^{i\gamma} \end{pmatrix} \tag{6.65}$$

Since the potential is real, if $\psi(x)$ is a solution then $\psi^*(x)$ is also a solution. So in addition to the solutions Eq. (6.50) we also have the following time reversed[5] solutions:

$$\psi_1(x) = \begin{cases} A^*e^{-ikx} + B^*e^{+ikx} & \text{for} \quad x < -a \\ C^*e^{-\kappa x} + D^*e^{+\kappa x} & \text{for} \quad -a < x < +a \\ F^*e^{-ikx} + G^*e^{+ikx} & \text{for} \quad +a < x \end{cases}. \tag{6.66}$$

Comparison with Eq. (6.59) reveals that effectively the directions of motion have been reversed and the coefficient A has been interchanged with B^*, and F with G^*. Hence we may make the following changes in Eq. (6.61)

$$A \longleftrightarrow B^* \quad \text{and} \quad F \longleftrightarrow G^*$$

to obtain equally valid equation

$$\begin{pmatrix} A^* \\ G^* \end{pmatrix} = \begin{pmatrix} S_{11} & S_{12} \\ S_{21} & S_{22} \end{pmatrix} \begin{pmatrix} B^* \\ F^* \end{pmatrix}. \tag{6.67}$$

Eq. (6.67) and Eq. (6.61) can be combined to yield the condition

$$S^*S = \hat{I}. \tag{6.68}$$

This condition together with te unitarity of S implies that S-matrix must be symmetric as a consequence of the time reversal symmetry.

Problem 6.6 From the Eq. (6.60) and Eq. (6.61) using the unitarity for a symmetric matrix
show that the matrix M has the following form

$$M = \begin{pmatrix} \frac{1}{S_{12}} & \frac{S_{11}^*}{S_{12}^*} \\ \frac{S_{11}}{S_{12}} & \frac{1}{S_{12}^*} \end{pmatrix}. \tag{6.69}$$

Also verify that $\det M = 1$.

---

[5]$t \to -t$ and complex conjugation

Since the potential is an even function of x, another solution is obtained by replacing x in Eq. (6.50) by $-x$

$$\psi_2(x) = \begin{cases} Ae^{-ikx} + Be^{+ikx} & \text{for} \quad x < -a \\ Ce^{+\kappa x} + De^{-\kappa x} & \text{for} \quad -a < x < +a \\ Fe^{-ikx} + Ge^{+ikx} & \text{for} \quad +a < x \end{cases} . \tag{6.70}$$

If Ge^{+ikx} is the wave incident on the barrier from the left, then Be^{+ikx} is the transmitted and Ae^{-ikx} is the incident from the right. Hence if we make the replacement $A \longleftrightarrow G$ and $B \longleftrightarrow F$ in Eq. (6.51) we obtain

$$\begin{pmatrix} F \\ B \end{pmatrix} = \begin{pmatrix} S_{11} & S_{12} \\ S_{21} & S_{22} \end{pmatrix} \begin{pmatrix} G \\ A \end{pmatrix}.$$

This relation can also be written as

$$\begin{pmatrix} B \\ F \end{pmatrix} = \begin{pmatrix} S_{22} & S_{21} \\ S_{12} & S_{11} \end{pmatrix} \begin{pmatrix} A \\ G \end{pmatrix}.$$

Hence invariance under reflection implies

$$S_{11} = S_{22} \quad \text{and} \quad S_{12} = S_{21}. \tag{6.71}$$

If the conservation of probability, time reversal and space reflection symmetries are to be valid simultaneously then the matrix M has to be of the following structure.

$$M_{11} = M_{22}^*, \ M_{12} = -M_{21}, \ \det M = +1.$$

We see that Eq. (6.57) and Eq. (6.58) are the results of very general properties, shared by all potential that are symmetric with respect to the origin and vanish at large values of $|x|$. For all such potentials that the solution of the Schrödinger equation must be asymptotically of the form

$$\psi(x) \sim \begin{cases} Ae^{+ikx} + Be^{-ikx} & \text{as} \quad x \to -\infty \\ Fe^{+ikx} + Ge^{-ikx} & \text{as} \quad x \to +\infty \end{cases} .$$

By virtue of the general arguments just presented, these two portions of the eigenfunctions are related by the matrix equation

$$\begin{pmatrix} A \\ B \end{pmatrix} = \begin{pmatrix} \alpha_1 + i\beta_1 & +i\beta_2 \\ -i\beta_2 & \alpha_1 - i\beta_1 \end{pmatrix} \begin{pmatrix} F \\ G \end{pmatrix},$$

with real parameters α_1, β_1 and β_2 subjrct to the additional constraint

$$\alpha_1^2 + \beta_1^2 - \beta_2^2 = 1.$$

The significance of the matrix method presented here is that it allows a clear separation beyween the initial conditions which can be adopted to suit any particular problem, and the matrices M and S which do not depend on any particular of the wave packet used. Once either of the matrices are worked out as a function of energy, all problems relating to the potential barrier have essentially been solved. For example, the transmission coefficien T is given by $\frac{|F|^2}{|A|^2}$ if $G = 0$, and therefore

$$T = \frac{1}{|M_{11}|^2} = |S_{21}|^2. \tag{6.72}$$

This work is in its most elementary form of the S-matrix theory of the more sophisticated scattering matrix theory of collisions.

6.4   Delta Function Potential

Consider a particle of mass m moving in an attractive δ-function potential

$$V(x) = -g\delta(x), \tag{6.73}$$

where g is a real positive constant. Integrating Eq. (6.73) one can verify that g has the dimension of energy $\times$ length, or equivalently it is square of the electrostatic charge. Though this is not a physical potential, but it serves as a useful *toy model* and is known as *one dimensional hydrogen atom*.

The Schrödinger equation for this potential is

$$-\frac{\hbar^2}{2m}\frac{d^2\psi(x)}{dx^2} - g\delta(x)\psi(x) = E\psi(x). \tag{6.74}$$

Integrating Eq. (6.74) from $-\epsilon$ to $+\epsilon$, and taking the limit $\epsilon \to 0$, we get

$$\lim_{\epsilon \to 0}\left[\left(\frac{d\psi}{dx}\right)_{x=+\epsilon} - \left(\frac{d\psi}{dx}\right)_{x=-\epsilon}\right] + \frac{2m}{\hbar^2}g\psi(0) = 0, \tag{6.75}$$

since

$$\int_{-\epsilon}^{+\epsilon} E\psi(x)\,dx \approx E\psi(0)\,2\epsilon \to 0, \quad \text{as } \epsilon \to 0.$$

This potential problem admits of both *bound* and *continuum* solutions.

(i). *Bound state solution* is given when $E < 0$ and the equation is

$$\frac{d^2\psi(x)}{dx^2} - \kappa^2\psi(x) = -\frac{2m}{\hbar^2}g\delta(x)\psi(x), \tag{6.76}$$

$$\kappa^2 = \frac{2m|E|}{\hbar^2}.$$

Since $\delta(x)$ is zero everywhere except at $x = 0$, $\psi(x)$ must satisfy

$$\frac{d^2\psi(x)}{dx^2} - \kappa^2\psi(x) = 0$$

everywhere except at $x = 0$. Assuming $\psi(x) \to 0$ at $x \to \pm\infty$, we have

$$\psi(x) = \begin{cases} Ae^{+\kappa x}, & \text{when} \quad x < 0, \\ Be^{-\kappa x} & \text{when} \quad x > 0 \end{cases} \tag{6.77}$$

Since $\psi(x)$ is continuous at $x = 0$, we have $A = B$, and

$$\psi(x) = \begin{cases} Ae^{+\kappa x}, & \text{when} \quad x \leq 0, \\ Ae^{-\kappa x} & \text{when} \quad x \geq 0 \end{cases} \tag{6.78}$$

Using Eq. (6.76) in Eq. (6.75),

$$-2\kappa A + \frac{2mAg}{\hbar^2} = 0, \quad \text{or} \quad \kappa = \frac{mg}{\hbar^2}$$

Since

$$\kappa = \sqrt{\frac{2m|E|}{\hbar^2}}, \quad \text{we have} \quad \frac{mg}{\hbar^2} = \sqrt{\frac{2m|E|}{\hbar^2}}$$

and the allowed energy is

$$E = -\frac{mg^2}{2\hbar^2}. \tag{6.79}$$

Energy is thus negative and there is only *one bound state* which can be normalized as follows

$$1 = \int_{-\infty}^{+\infty} |\psi(x)|^2 dx = 2|A|^2 \int_0^{+\infty} e^{-2\kappa x} dx \tag{6.80}$$
$$\text{or } |A|^2 = \kappa \quad i.e. \quad A = \sqrt{\kappa}.$$

(ii). For *continuum solutions* we have $E > 0$. We assume that a particle is incident from the left. The solutions may be written as

$$\psi(x) = \begin{cases} e^{+ikx} + re^{-ikx} & \text{when} \quad x < 0 \\ te^{+ikx} & \text{when} \quad x > 0 \end{cases} \tag{6.81}$$

We have already chosen the normalization so that the coefficient of the first term is one. Here $k = \sqrt{2mE}/\hbar$. Then the continuity at $x = 0$ and Eq. (6.75) give

$$1 + r = \psi\left(0^-\right) = \psi\left(0^+\right) = t, \tag{6.82}$$

$$\text{and} \quad \left(\frac{d\psi}{dx}\right)_{0^+} - \left(\frac{d\psi}{dx}\right)_{0^-} = -\frac{2mg}{\hbar^2}\psi\left(0\right),$$

$$\text{or} \quad ik\left(t - 1 + r\right) = -\frac{2mg}{\hbar^2}\psi\left(0\right) = -\frac{2mg}{\hbar^2}\left(1 + r\right), \tag{6.83}$$

$$\text{whence} \quad r = \frac{i\frac{\alpha}{k}}{1 - i\frac{\alpha}{k}},$$

$$\text{and} \quad t = \frac{1}{1 - i\frac{\alpha}{k}},$$

$$\text{with} \quad \alpha = \frac{mg}{\hbar^2}.$$

The reflection and the transmission coefficients are

$$R = |r|^2 = \frac{\frac{\alpha^2}{k^2}}{1 + \frac{\alpha^2}{k^2}} = \frac{\alpha^2}{\alpha^2 + k^2} = \frac{mg^2}{mg^2 + 2E\hbar^2},$$

$$T = |t|^2 = \frac{1}{1 + \frac{\alpha^2}{k^2}} = \frac{k^2}{\alpha^2 + k^2} = \frac{2E\hbar^2}{mg^2 + 2E\hbar^2},$$

and we have

$$R + T = 1. \tag{6.84}$$

6.5 Oscillator Problem by Schrödinger Method

Linear harmonic oscillator is one of the very few problems in quantum mechanics which can be solved *exactly*. Its importance can hardly be over-emphasized. It has got application in almost *all* branches of physics. From condense matter physics to spectroscopy, from nuclear physics to quantum field theory we have to apply linear harmonic oscillator in one form or other.

The Hamiltonian is given by

$$H = -\frac{\hbar^2}{2m}\frac{d^2}{dx^2} + \frac{1}{2}m\omega^2 x^2, \tag{6.85}$$

where $\omega = \sqrt{\kappa/m}$, κ being the spring constant. And the Schrödinger equation is

$$\left[\frac{d^2}{dx^2} - \frac{\omega^2 m^2 x^2}{\hbar^2} + \frac{2mE}{\hbar^2}\right] \psi(x) = 0. \tag{6.86}$$

Changing over to dimensionless quantities

$$\xi = x\sqrt{\frac{m\omega}{\hbar}}, \quad \epsilon = \frac{2E}{\hbar\omega}, \tag{6.87}$$

we get the second order equation

$$\left[\frac{d^2}{d\xi^2} - \xi^2 + \epsilon\right] \psi(x) = 0. \tag{6.88}$$

We study the asymptotic behaviour of $\psi(x)$ as $\xi \to \infty$, when

$$\left[\frac{d^2}{d\xi^2} - \xi^2\right] \psi_\infty(x) = 0.$$

Thus for $|\xi| \to \infty$ the wavefunction must have the form $\lim_{|\xi|\to\infty} \psi(\xi) = \exp\left(\pm\frac{\xi^2}{2}\right)$. Of the two solutions we retain $\exp\left(-\frac{\xi^2}{2}\right)$, as the other one blows up as $\xi \to \infty$. Then we seek a solution of Eq. (6.88) in the form

$$\psi(\xi) = v(\xi)\exp\left(-\frac{\xi^2}{2}\right). \tag{6.89}$$

Substituting Eq. (6.89) in Eq. (6.88) we get

$$v'' - 2\xi v' + (\epsilon - 1)v = 0, \tag{6.90}$$

where *primes* indicate differentiation with respect to ξ. In order that Eq. (6.89) is finite for $\xi \to \infty$, it is necessary that the solutions v should be polynomial of finite order in ξ. Considering the power series solution of Eq. (6.90) of the form

$$v = \sum_{k=0}^{\infty} a_k \xi^k, \tag{6.91}$$

it is possible to show that unless the series terminates, *i.e.* unless $a_k = 0$ for all $k > n$ where n is some integer, the function v will approach infinity more rapidly than $\exp(\xi^2)$ and hence $\psi = v\exp(-\xi^2/2)$ will be infinite as $\xi \to \infty$. We can verify this by substituting Eq. (6.91) in Eq. (6.90) and then equating the coefficients of like powers of ξ to zero. We get the following recursion relationships for the coefficients a_k.

$$a_{k+2} = a_k \frac{2k - (\epsilon - 1)}{(k+2)(k+1)}. \tag{6.92}$$

Thus once a_0 and a_1 are specified (and the two numbers may be chosen independently) we can determine the rest of the coefficients from this relation. Now $\frac{a_{k+2}}{a_k} \rightarrow \frac{2}{k}$ as $k \rightarrow \infty$. Comparing this with the limiting ratio $\frac{1}{k}$ of two successive terms in the expansion of

$$\xi^\gamma \exp\left(\xi^2\right) = \sum_{k=0}^{\infty} \frac{1}{k!} \xi^{\gamma+2k},$$

where γ is arbitrary, we conclude that unless the series for $v(\xi)$ terminates, $v(\xi)$ will approach infinity more rapidly than $\xi^\gamma \exp(\xi^2)$ and $\psi(\xi)$ will approach infinity like $\exp(\xi^2/2)$. Substituting $k = n$ into the recursion relation Eq. (6.92), we find that

$$\epsilon - 1 = 2n, \quad n = 0, 1, 2, \cdots \tag{6.93}$$

is the condition that the series will terminate at $k = n$.

To each value of n in Eq. (6.93), there corresponds a polynomial of order n, which is called a *Hermite polynomial*,

$$H_n(\xi) = (-1)^n \exp\left(\xi^2\right) \frac{d^n}{d\xi^n} \exp\left(-\xi^2\right). \tag{6.94}$$

Thus the solution of Eq. (6.90), in the form of a polynomial of finite degree exact up to a normalization constant N, is

$$v_n(\xi) = N_n H_n(\xi). \tag{6.95}$$

Then the stationary states of a linear harmonic oscilletor is

$$\psi_n(\xi) = N_n \exp\left(-\frac{\xi^2}{2}\right) H_n(\xi), \tag{6.96}$$

with
$$N_n = \left(\sqrt{\pi} n! 2^n\right)^{-\frac{1}{2}},$$

and
$$\int_{-\infty}^{+\infty} \psi_n(\xi) \psi_m(\xi) d\xi = \delta_{n,m}.$$

To each state which is represented by a wavefunction of the form Eq. (6.96) there corresponds according to Eq. (6.93), one value $\epsilon_n = 2n + 1$ (non-degerate). Using Eq. (6.87) we find the value of the energy of a harmonic oscillator

$$E = \hbar\omega\left(n + \frac{1}{2}\right). \tag{6.97}$$

The energy of the *ground state*

$$E_0 \; = \; \frac{1}{2}\hbar\omega$$

is called the *zero point energy*. Since the potential energy of the oscillator is invariant with respect to the parity operation $\xi \to -\xi$, the stationary states can be divided into *even* and *odd* parity, according as n is even or odd respectively.

Even Parity   Odd Parity

$$H_0\left(\xi\right) \; = \; 1 \qquad\qquad\qquad H_1\left(\xi\right) \; = \; 2\xi$$
$$H_2\left(\xi\right) \; = \; 4\xi^2 - 2 \qquad\qquad H_3\left(\xi\right) \; = \; 8\xi^3 - 12\xi$$
$$H_4\left(\xi\right) \; = \; 16\xi^4 - 48\xi^2 + 12 \quad \cdots$$

In the general case, the parity of the wavefunction is determined from Eq. (6.94).
    The Hermite polynomials satisfy the simple recursion relationships

$$\xi H_n\left(\xi\right) \; = \; nH_{n-1}\left(\xi\right) + \frac{1}{2}H_{n+1}\left(\xi\right), \tag{6.98}$$

$$\frac{dH_n\left(\xi\right)}{d\xi} \; = \; 2nH_{n-1}\left(\xi\right). \tag{6.99}$$

These formulas are very useful in actual calculations. The mean value of ξ in the nth state is

$$\langle\xi\rangle_n \; = \; \int_{-\infty}^{+\infty} \psi_n^2\left(\xi\right) \cdot \xi \, d\xi \; = \; 0$$

since te integrand is an odd function of ξ. Thus

$$\langle(\Delta\xi)^2\rangle_n \; = \; \langle\xi^2\rangle_n - \langle\xi\rangle_n^2 \; = \; \langle\xi^2\rangle_n$$
$$= \; \int_{-\infty}^{+\infty} \psi_n\left(\xi\right)\xi^2\psi_n\left(\xi\right)d\xi. \tag{6.100}$$

Using Eq. (6.96) and Eq. (6.98), we find

$$\xi\psi_n \; = \; \sqrt{\frac{n}{2}}\psi_{n-1} + \sqrt{\frac{n+1}{2}}\psi_{n+1}, \tag{6.101}$$

and hence

$$\xi^2\psi_n \; = \; \frac{1}{2}\sqrt{n\left(n-1\right)}\psi_{n-2} + \left(n+\frac{1}{2}\right)\psi_n + \frac{1}{2}\sqrt{\left(n+1\right)\left(n+2\right)}\psi_{n+2}. \tag{6.102}$$

Substituting Eq. (6.102) in Eq. (6.100) and taking into account the orthonormality of the wavefunctions we get

$$\langle \xi^2 \rangle_n = n + \frac{1}{2}, \quad \text{or} \quad \langle x^2 \rangle_n = \left(n + \frac{1}{2} \right) \frac{\hbar}{m\omega} , \tag{6.103}$$

or mean value of x^2 in $n = 0$ ground state is

$$\langle x^2 \rangle_0 = \frac{\hbar}{2m\omega} .$$

6.6 Linear Harmonic Oscillator by Operator Method

We shall now develop Dirac's operator method to solve the harmonic oscillator problem. The Hamiltonian of the linear harmonic oscillator is

$$H = \frac{\hat{p_x}^2}{2m} + \frac{1}{2}m\omega^2 x^2, \tag{6.104}$$

where ω is the angular frequency of the classical oscillator which is related to the spring constant K by the relation $\omega = \sqrt{K/m}$.

We now define two *non-hermitian* operators

$$\hat{a}_\pm = \mp \frac{i}{\sqrt{2}} \left[\frac{\hat{p_x}}{\sqrt{m\hbar\omega}} \pm i\sqrt{\frac{m\omega}{\hbar}} \hat{x} \right] . \tag{6.105}$$

Problem 6.7 *(i)*. Check that the operators in Eq. (6.105) are dimensionless.

(ii). Also show that $\hat{a}_+$ and $\hat{a}_-$ are Hermitian adjoints of each other.

$$\hat{a}_+^\dagger = \hat{a}_-, \quad \hat{a}_-^\dagger = \hat{a}_+. \tag{6.106}$$

(iii). Using the basic commutation relation

$$[\hat{x}, \hat{p_x}] = i\hbar,$$

show that

$$[\hat{a}_-, \hat{a}_+] = \hat{I} . \tag{6.107}$$

Now we express $\hat{x}$ and $\hat{p}_x$ in terms of $\hat{a}_+$ and $\hat{a}_-$:

$$\hat{x} = \sqrt{\frac{\hbar}{2m\omega}}\left(\hat{a}_+ + \hat{a}_-\right),$$
$$\hat{p}_x = i\sqrt{\frac{m\hbar\omega}{2}}\left(\hat{a}_+ - \hat{a}_-\right) \tag{6.108}$$

and the Hamiltonian is

$$\hat{H} = \frac{\hbar\omega}{2}\left(\hat{a}_+\hat{a}_- + \hat{a}_-\hat{a}_+\right). \tag{6.109}$$

Problem 6.8 Using the commutation relation Eq. (6.107), show that

$$\hat{H} = \hbar\omega\left(\hat{a}_+\hat{a}_- + \frac{1}{2}\right). \tag{6.110}$$

We define the *number operator* $\hat{N}$ as

$$\hat{N} = \hat{a}_+\hat{a}_-, \tag{6.111}$$

so that

$$\hat{H} = \hbar\omega\left(\hat{N} + \frac{1}{2}\right). \tag{6.112}$$

Since

$$\hat{N}^\dagger = \hat{a}_-^\dagger\hat{a}_+^\dagger = \hat{a}_+\hat{a}_- = \hat{N}, \tag{6.113}$$

the number operator $\hat{N}$ is *Hermitian*.

Problem 6.9 Show that

$$\left[\hat{H}, \hat{a}_\pm\right] = \pm\hbar\omega\hat{a}_\pm. \tag{6.114}$$

Now the eigenvalue equation of $\hat{H}$ is

$$\hat{H}|E\rangle = E|E\rangle. \tag{6.115}$$

From Eq. (6.114)

$$\left[\hat{H}, \hat{a}_\pm\right]|E\rangle = \pm\hbar\omega\hat{a}_\pm|E\rangle,$$
$$\left[\hat{H}\hat{a}_\pm - \hat{a}_\pm\hat{H}\right]|E\rangle = \pm\hbar\omega\hat{a}_\pm|E\rangle, \tag{6.116}$$
$$\hat{H}\left[\hat{a}_\pm|E\rangle\right] = \hat{a}_\pm\hat{H}|E\rangle \pm \hbar\omega\hat{a}_\pm|E\rangle$$
$$= \left(E \pm \hbar\omega\right)\left[\hat{a}_\pm|E\rangle\right]. \tag{6.117}$$

From Eq. (6.117) we notice that the kets $\hat{a}_{\pm}|E\rangle$ are also eigenkets of $\hat{H}$ belonging to the eigenvalues $(E \pm \hbar\omega)$.

Thus $\hat{a}_+$ and $\hat{a}_-$ operating on $|E\rangle$ respectively raises and lowers the eigenvalue E by $\hbar\omega$ and are called *raising* and *lowering* operators[6] Since $\hat{H}$ contains squares of the Hermitian operators $\hat{p}_x$ and $\hat{x}$, the expectation value of $\hat{H}$ in any of its eigenkets cannot be negative and its eigenvalues are thus non-negative. If E_0 is the smallest of the eigenvalues of $\hat{H}$ and $|E_0\rangle$ is the corresponding eigenket, we must have

$$\hat{a}_-|E_0\rangle = 0. \tag{6.118}$$

For otherwise $\hat{a}_-|E_0\rangle$ would be an eigenket of $\hat{H}$ with values $E_0 - \hbar\omega$, which is lower than E_0, contrary to the assumption that E_0 is the lowest allowed eigenvalue.

Now from Eq. (6.110) and Eq. (6.118)

$$
\begin{aligned}
(\hat{a}_+\hat{a}_-)|E_0\rangle &= \hat{N}|E_0\rangle = \left(\hat{H} - \frac{\hbar\omega}{2}\right)|E_0\rangle \\
&= \left(E_0 - \frac{\hbar\omega}{2}\right)|E_0\rangle.
\end{aligned}
\tag{6.119}
$$

Since the left hand side is 0, we get

$$E_0 = \frac{\hbar\omega}{2}, \tag{6.120}$$

i.e. the lowest energy eigenvalue of the oscillator is $E_0 = \frac{\hbar\omega}{2}$, which is the ground state energy.

We can now operate on $|E_0\rangle$ repeatedly with the raising operator $\hat{a}_+$ and obtain the sequence of eigenkets

$$|E_0\rangle, \ \hat{a}_+|E_0\rangle, \ \hat{a}_+^2|E_0\rangle, \ \cdots \tag{6.121}$$

which are *not* normalized. The eigenket $\hat{a}_+^n|E_0\rangle$ has eigenvalues

$$E_n = \left(n + \frac{1}{2}\right)\hbar\omega, \quad \text{with} \quad n = 0, 1, 2, \cdots, \tag{6.122}$$

positive integers starting with 0.

If $|E_n\rangle$ be the normalized eigenket corresponding to the eigenvalue E_n and $|E_{n+1}\rangle$ be that corresponding to the eigenvalue E_{n+1}, then from Eq. (6.121) we can write

$$|E_{n+1}\rangle = c_{n+1}\hat{a}_+|E_n\rangle, \tag{6.123}$$

---

[6]In quantum field theoretic terminology they are called *creation* operator $\hat{a}^\dagger$ and *annihilation* or *destruction* oparor $\hat{a}$ respectively.

where c_{n+1} is the normalization constant.

$$\langle E_{n+1}|E_{n+1}\rangle = |c_{n+1}|^2 \langle E_n|\hat{a}_+^\dagger \hat{a}_+|E_n\rangle = 1, \tag{6.124}$$

and since $\hat{a}_+^\dagger = \hat{a}_-$ from Eq. (6.106), therefore

$$|c_{n+1}|^2 \langle E_n|\hat{a}_- \hat{a}_+|E_n\rangle = 1. \tag{6.125}$$

Now

$$\hat{a}_- \hat{a}_+ = \frac{\hat{H}}{\hbar\omega} + \frac{1}{2}$$

$$\text{and } \hat{H}|E_n\rangle = E_n|E_n\rangle = \left(n + \frac{1}{2}\right)\hbar\omega|E_n\rangle,$$

$$\therefore \ \hat{a}_- \hat{a}_+|E_n\rangle = (n+1)|E_n\rangle. \tag{6.126}$$

Then from Eq. (6.125) and Eq. (6.126) we have

$$(n+1)|c_{n+1}|^2 = 1. \tag{6.127}$$

Taking c_{n+1} to be real positive, we have

$$c_{n+1} = \frac{1}{\sqrt{n+1}}. \tag{6.128}$$

ENERGY REPRESENTATION OF LINEAR HARMONIC OSCILLATOR

When the orthonormal complete set of eigenkets $\{|E_n\rangle\}$ is taken as the basis vectors of the oscillator ket space then the representation is called *energy representation*. This is a discrete representation and the matrix forms of $\hat{H}$ and $\hat{N}$ are obviously diagonal in this representation:

$$\hat{H} = \begin{pmatrix} \frac{1}{2} & 0 & 0 & \cdots \\ 0 & \frac{3}{2} & 0 & \cdots \\ 0 & 0 & \frac{5}{2} & \cdots \\ \vdots & \vdots & \vdots & \ddots \end{pmatrix} \quad \text{and} \quad \hat{N} = \begin{pmatrix} 0 & 0 & 0 & \cdots \\ 0 & 1 & 0 & \cdots \\ 0 & 0 & 2 & \cdots \\ \vdots & \vdots & \vdots & \ddots \end{pmatrix} \tag{6.129}$$

Problem 6.10 Use Eq. (6.123) and Eq. (6.128), also orthogonality of energy eigenkets and check the following forms of the matrices.

$$\hat{a}_+ = \begin{pmatrix} 0 & 0 & 0 & \cdots \\ \sqrt{1} & 0 & 0 & \cdots \\ 0 & \sqrt{2} & 0 & \cdots \\ \vdots & \vdots & \vdots & \ddots \end{pmatrix} \quad \text{and} \quad \hat{a}_- = \begin{pmatrix} 0 & \sqrt{1} & 0 & \cdots \\ 0 & 0 & \sqrt{2} & \cdots \\ 0 & 0 & 0 & \cdots \\ \vdots & \vdots & \vdots & \ddots \end{pmatrix} \tag{6.130}$$

(i) Time Evolution of Oscillator

From the Heisenberg equation of motion Eq. (4.47) we can write the following equations for $\hat{p}_x$ and $\hat{x}$ of the linear harmonic oscillator.

$$\frac{d\hat{p}_x}{dt} = \frac{1}{i\hbar}\left[\hat{p}_x, \frac{\hat{p}_x^2}{2m} + \frac{1}{2}m\omega^2\hat{x}^2\right]$$
$$= -m\omega^2\hat{x}, \qquad \text{and} \tag{6.131}$$
$$\frac{d\hat{x}}{dt} = \frac{1}{i\hbar}\left[\hat{x}, \frac{\hat{p}_x^2}{2m} + \frac{1}{2}m\omega^2\hat{x}^2\right]$$
$$= \frac{\hat{p}_x}{m}. \tag{6.132}$$

Since these are coupled equations we consider the time dependence of $\hat{a}_+$ and $\hat{a}_-$:

$$\frac{d\hat{a}_-}{dt} = -i\omega\hat{a}_-, \qquad \text{and} \tag{6.133}$$

$$\frac{d\hat{a}_+}{dt} = +i\omega\hat{a}_+, \tag{6.134}$$

whose solutions are

$$\hat{a}_-(t) = \hat{a}_-(0)\exp(-i\omega t), \qquad \text{and} \tag{6.135}$$
$$\hat{a}_+(t) = \hat{a}_+(0)\exp(+i\omega t). \tag{6.136}$$

Thus from Eq. (6.105)

$$\frac{\hat{p}_x(t)}{\sqrt{m\hbar\omega}} + i\sqrt{\frac{m\omega}{\hbar}}\hat{x}(t) = \frac{\hat{p}_x(0)}{\sqrt{m\hbar\omega}}e^{+i\omega t} + i\sqrt{\frac{m\omega}{\hbar}}\hat{x}(0)e^{+i\omega t}, \tag{6.137}$$

and

$$\frac{\hat{p}_x(t)}{\sqrt{m\hbar\omega}} - i\sqrt{\frac{m\omega}{\hbar}}\hat{x}(t) = \frac{\hat{p}_x(0)}{\sqrt{m\hbar\omega}}e^{-i\omega t} - i\sqrt{\frac{m\omega}{\hbar}}\hat{x}(0)e^{-i\omega t}. \tag{6.138}$$

Eq. (6.137) is the Hermitian conjugate of Eq. (6.138). Equating the Hermitian and the anti-Hermitian term in either equation we get

$$\hat{p}_x(t) = \hat{p}_x(0)\cos\omega t - m\omega\hat{x}(0)\sin\omega t, \qquad \text{and} \tag{6.139}$$

$$\hat{x}(t) = \hat{x}(0)\cos\omega t + \frac{\hat{p}_x(0)}{m\omega}\sin\omega t. \tag{6.140}$$

These equations are analogous to the classical equations of motions, showing $\hat{x}(t)$, $\hat{p}_x(t)$ oscillating with the angular frequancy ω.

(ii) Coherent State

Study of *coherent states* was pioneered by R. Glauber. We indicate some of the features in this subsection.

From the expressions Eq. (6.139) and Eq. (6.140) one might conclude that $\langle \hat{x} \rangle$ and $\langle \hat{p}_x \rangle$ always oscillate with angular frequency ω. However, that is *not* correct. For example the expectation value $\langle n | \hat{x}(t) | n \rangle$ with energy eigenkets $|n\rangle$ vanishes, because $\hat{x}(0)$ and $\hat{p}_x(0)$ change n by ± 1 and the states $|n\rangle$ and $|n\pm 1\rangle$ are orthogonal. To observe oscillations as in a classical oscillator we need to take a superposition of energy eigenkets such as

$$|\alpha\rangle \;=\; c_0 |0\rangle + c_1 |1\rangle. \tag{6.141}$$

Then the expectation value $\langle \alpha | \hat{x} | \alpha \rangle$ does oscillate instead of being zero.

We now ask how can we construct a superposition of energy eigenkets that most closely resembles the classical oscillator. In other words, we want a wavepacket that bounces back and forth without spreading in shape. A *coherent state* is such a wavepacket. It is defined as the eigenfunction of the non-Hermitian *annihilation operator* $\hat{a}_-$

$$\hat{a}_- |\alpha\rangle \;=\; \alpha |\alpha\rangle, \tag{6.142}$$

where α is a complex number in general.

We now prove a very useful identity

$$e^{\hat{A}} \hat{B} e^{-\hat{A}} \;=\; B + \left[\hat{A}, \hat{B} \right] + \frac{1}{2!} \left[\hat{A}, \left[\hat{A}, \hat{B} \right] \right] + \frac{1}{3!} \left[\hat{A}, \left[\hat{A}, \left[\hat{A}, \hat{B} \right] \right] \right] + \cdots \tag{6.143}$$

Proof:

Let $f(\lambda) = e^{\lambda \hat{A}} B e^{-\lambda \hat{A}}$. Making a Taylor Series expansion of $f(\lambda)$ and noting that

$$\begin{aligned}
\frac{df(\lambda)}{d\lambda} &= \hat{A} f(\lambda) - f(\lambda) \hat{A} = \left[\hat{A}, f(\lambda) \right], \\
\frac{d^2 f(\lambda)}{d\lambda^2} &= \left[\hat{A}, \frac{df(\lambda)}{d\lambda} \right] = \left[\hat{A}, \left[\hat{A}, f(\lambda) \right] \right], \\
\cdots &= \cdots ,
\end{aligned}$$

and since $f(0) = \hat{B}$, we get

$$f(\lambda) \;=\; \hat{B} + \frac{\lambda}{1!} \left[\hat{A}.\hat{B} \right] + \frac{\lambda^2}{2!} \left[\hat{A}, \left[\hat{A}, \hat{B} \right] \right] + \cdots .$$

from which Eq. (6.143) follows by setting $\lambda = 1$.

Similarly it can be shown, if $\hat{A}$ and $\hat{B}$ are two operators both of which commute with $\left[\hat{A}, \hat{B}\right]$,

$$e^{\hat{A}} e^{\hat{B}} = e^{\hat{A}+\hat{B}+\frac{1}{2}[\hat{A},\hat{B}]}. \tag{6.144}$$

We consider the unitary operator $\hat{S}_\alpha = \exp\left(\alpha \hat{a}_+ + \alpha^* \hat{a}_-\right)$ and using Eq. (6.143) we obtain

$$\exp\left(\alpha \hat{a}_+ - \alpha^* \hat{a}_-\right) \hat{a}_- \exp\left(-\alpha \hat{a}_+ + \alpha^* \hat{a}_-\right) = \hat{a}_- - \alpha \hat{I}. \tag{6.145}$$

Problem 6.11 Prove Eq. (6.145).

The unitary operator S_α has the following properties

$$\hat{S}_\alpha^\dagger = \hat{S}_\alpha^{-1} = \hat{S}_{-\alpha}.$$

From Eq. (6.145)

$$\begin{aligned}
\hat{S}_\alpha \hat{a}_- \hat{S}_\alpha^{-1} &= \hat{a}_- - \alpha \hat{I}, \\
\hat{S}_\alpha \hat{a}_- &= \hat{a}_- \hat{S}_\alpha - \alpha \hat{S}_\alpha, \\
\hat{S}_\alpha \hat{a}_- |0\rangle &= \hat{a}_- \hat{S}_\alpha |0\rangle - \alpha \hat{S}_\alpha |0\rangle,
\end{aligned} \tag{6.146}$$

$$a_- \left[S_\alpha |0\rangle \right] = \alpha \left[s_{alpha} |0\rangle \right]. \tag{6.147}$$

where $|0\rangle$ is the oscillator ground state. Thus $S_\alpha |0\rangle$ is the eigenket of a_- with eigenvalue α.

$$\therefore \ \hat{S}_\alpha |0\rangle = |\alpha\rangle$$

is called the *coherent state*.

Since

$$|\alpha\rangle = \hat{S}_\alpha |0\rangle = \exp\left(\alpha \hat{a}_+ - \alpha^* \hat{a}_-\right) |0\rangle, \tag{6.148}$$

using Eq. (6.143) and Eq. (6.145) one can show that

$$\begin{aligned}
|\alpha\rangle &= \exp\left(\alpha \hat{a}_+ - \alpha^* \hat{a}_-\right) |0\rangle \\
&= \exp\left(-\frac{|\alpha|^2}{2}\right) \exp\left(\alpha \hat{a}_+\right) |0\rangle \tag{6.149} \\
&= \exp\left(-\frac{|\alpha|^2}{2}\right) \sum_{n=0}^{\infty} \frac{\alpha^n}{n!} \hat{a}_+^n |0\rangle \\
&= \exp\left(-\frac{|\alpha|^2}{2}\right) \sum_{n=0}^{\infty} \frac{\alpha^n}{\sqrt{n!}} |n\rangle. \tag{6.150}
\end{aligned}$$

Thus the probability of finding the nth state is given by

$$|\langle n|\alpha\rangle|^2 \;=\; P_n \;\equiv\; e^{-|\alpha|^2}\frac{\alpha^{2n}}{\sqrt{n!}} \tag{6.151}$$

which is a *Poisson distribution*.

Again we have

$$
\begin{aligned}
\hat{a}_-|\alpha\rangle &= \exp\left(-\frac{|\alpha|^2}{2}\right)\sum_{n=0}^{\infty}\frac{\alpha^n}{\sqrt{n!}}\hat{a}_-|n\rangle \\
&= \exp\left(-\frac{|\alpha|^2}{2}\right)\sum_{n=1}^{\infty}\frac{\alpha^n}{\sqrt{n!}}\sqrt{n}|n-1\rangle \\
&= \exp\left(-\frac{|\alpha|^2}{2}\right)\alpha\sum_{n=0}^{\infty}\frac{\alpha^n}{\sqrt{n!}}|n\rangle \\
&= \alpha|\alpha\rangle.
\end{aligned}
$$

We shall add a note here about the notation used. The lowering operator a_- and the raising operator a_+ has been used remembering that they change the state of the system from n to $n-1$ or to $n+1$ respectively. In all these transition a field quanta (phonon or photon) are created or destroyed (annihilated) respectively. That is why in field theory the corresponding operators are called *creation operator* $a^\dagger$ or *destruction (annihilation) operator* a respectively.

6.7   Periodic Potential

This is the final example of one dimensional problems where we consider motion of a particle in a periodic potential illustrated in Fig. (6.7). This is called the Kronig-Penny potential which can be used as a *model* of interaction to which electrons and ions are subjected in a crystal lattice consisting of a regular array of single atoms separated by the distance l.

We have

$$V(x\pm l) \;=\; V(x).$$

We use the unitary translation operator $\mathcal{T}$ defined in Chapter 3 and we have

$$
\begin{aligned}
\hat{\mathcal{T}}^\dagger(l)\,\hat{x}\hat{\mathcal{T}}(l) &= x+l, \quad\text{and} \\
\hat{\mathcal{T}}(l)|x'\rangle &= |x'+l\rangle, \quad\text{and} \tag{6.152}\\
\hat{\mathcal{T}}^\dagger(l)\,V(x)\,\hat{\mathcal{T}}(l) &= V(x+l) \;=\; V(x). \tag{6.153}
\end{aligned}
$$

Kronig–Penny Potential

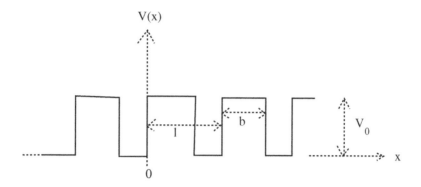

Figure 6.7: The Kronig-Penny Model of periodic potential

Since the kinetic energy does not change under translation, the Hamiltonian does not change under specific translation l, though it is not invariant under a general translation y. Thus we have

$$\hat{T}^\dagger(l)\,\hat{H}\hat{T}(l) \;=\; \hat{H}. \tag{6.154}$$

Since $\hat{T}(l)$ is unitary, *i.e.* $\hat{T}^\dagger(l) = \hat{T}(l)^{-1}$, we get

$$\hat{H}\hat{T}(l) \;=\; \hat{T}(l)\,\hat{H}, \tag{6.155}$$

i.e. $\hat{H}$ commutes with $\hat{T}(l)$ and they have simultaneous eigenfunctions. Since $\hat{T}$ is unitary but not Hermitian its eigenvalue is a complex number with modulus unity. If the barrier height between lattice sites is infinity the particle in the lattice is completely localized as there is no penetration or tunneling. The ground state in such a potential is one where the particle is localized at the n-th site say, $|n\rangle$, which is eigenstate of $\hat{H}$ belonging to the ground state energy E_0, *i.e.*

$$\hat{H}\,|n\rangle \;=\; E_0\,|n\rangle. \tag{6.156}$$

The particle wavefunction $\langle x'|n\rangle$ is finite only in the n-th site. Similar states localized at other sites also has the same energy. (n is the designation of the site not of energy level.) So there are infinite number of ground states n, where n varies from $-\infty$ to $+\infty$. Since

$$\hat{T}(l)\,|n\rangle \;=\; |n+1\rangle, \tag{6.157}$$

we observe that $|n\rangle$ is not an eigenstate of $\hat{\mathcal{T}}(l)$, though it commutes with $\hat{H}$. Thus there is an infinite-fold degeneracy and we have to construct the simultaneous eigenket of $\hat{H}$ and $\hat{\mathcal{T}}(l)$ by taking a linear combination of all the degerate kets $|n\rangle$ with n varying from $-\infty$ to $+\infty$.

We define

$$|\Psi\rangle = \sum_{n=-\infty}^{+\infty} e^{in\theta}|n\rangle, \tag{6.158}$$

where θ is a real parameter in the range $-\pi \le \theta \le +\pi$. We now demand $|\Psi\rangle$ to be simultaneous eigenket of $\hat{H}$ and $\hat{\mathcal{T}}(l)$:

$$\hat{H}|\Psi\rangle = E_0 \sum_{n=-\infty}^{+\infty} e^{in\theta}|n\rangle = E_0|\Psi\rangle, \tag{6.159}$$

$$\begin{aligned}
\hat{\mathcal{T}}(l)|\Psi\rangle &= \sum_{n=-\infty}^{+\infty} e^{in\theta}|n+1\rangle \\
&= \sum_{n=-\infty}^{+\infty} e^{i(n-1)\theta}|n\rangle = e^{-i\theta} \sum_{n=-\infty}^{+\infty} e^{in\theta}|n\rangle \\
&= e^{-i\theta}|\Psi\rangle.
\end{aligned} \tag{6.160}$$

Thus $|\Psi\rangle$ is so constructed that it is an eigenket of $\hat{\mathcal{T}}(l)$ with an eigenvalue which is a complex number with modulus 1. $|\Psi\rangle$ is thus parametrized by a continuous parameter θ.

In a real situation barriers between adjacent sites are not infinitely high and there will be quantum tunneling and therefore there will be leakage of wavefunction into neighbouring sites. In that case the Hamiltonian is not strictly diagonal although we have $\langle n|\hat{H}|n\rangle = E_0$ and all the diagonal elements are equal. One may then assume nearest neighbour interaction used in solid state physics, which means that one assumes that the non-diagonal elements of $\hat{H}$ are non-zero only for immediate neighbours, i.e.

$$\langle n'|\hat{H}|n\rangle \neq 0, \quad \text{only for } n' = n, n \pm 1. \tag{6.161}$$

We define

$$\langle n \pm 1|\hat{H}|n\rangle = -\Delta, \tag{6.162}$$

which is independent of n because of translational symmetry of $\hat{H}$. Since $|n\rangle$ and $|n'\rangle$ are orthogonal when $n \neq n'$, we get

$$\hat{H}|n\rangle = E_0|n\rangle - \Delta|n+1\rangle - \Delta|n-1\rangle. \tag{6.163}$$

$|n\rangle$ is no longer an eigenket of $\hat{H}$.

We again construct a linear combination

$$|\Psi\rangle = \sum_{n=-\infty}^{+\infty} e^{in\theta}|n\rangle. \tag{6.164}$$

$|\Psi\rangle$ is clearly an eigenket of $\hat{\mathcal{T}}(l)$. To see whether $|\Psi\rangle$ is an energy eigenket, we operate $\hat{H}$ on it:

$$
\begin{aligned}
\hat{H}|\Psi\rangle &= \sum_{n=-\infty}^{+\infty} e^{in\theta} \hat{H}|n\rangle \\
&= E_0 \sum_{n=-\infty}^{+\infty} e^{in\theta}|n\rangle - \Delta \sum_{n=-\infty}^{+\infty} e^{in\theta}|n+1\rangle - \Delta \sum_{n=-\infty}^{+\infty} e^{in\theta}|n-1\rangle \\
&= E_0 \sum_{n=-\infty}^{+\infty} e^{in\theta}|n\rangle - \Delta \sum_{n=-\infty}^{+\infty} \left(e^{-i\theta} + e^{+i\theta}\right) e^{in\theta}|n\rangle \\
&= \left(E_0 - 2\Delta\cos\theta\right)|\Psi\rangle. \tag{6.165}
\end{aligned}
$$

The energy eigenvalue now depends on the continuous parameter θ. The degeneracy is thus lifted as Δ is finite and we have a continuous *band* of energy eigenvalues between $E_0 - 2\Delta$ and $E_0 + 2\Delta$.

The energy eigenvalue equation is independent of the detailed shape of the potential as long as the nearest neighbour approximation is valid. As a result of tunneling the denumerably infinite fold degeneracy is now completely lifted and the allowed energy values form a continuous *band* between $E_0 \pm 2\Delta$, known as the Brillouin zone. In Fig. (6.8) we show schematically the formation of energy bands in the presence of crystal potential.

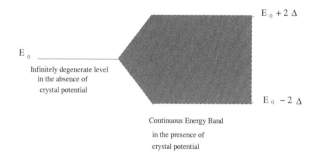

Figure 6.8: Formation of energy band in the presence of crystal potential.

Chapter 7

Rotation and Angular Momentum

7.1 Introduction

The angular momentum operator is closely related to rotation in space. The quantum mechanical definition of angular momentum differs from the classical definition $\mathbf{r} \times \mathbf{p}$ of angular momentum. This expression is *not* sufficiently general as it cannot yield spin angular momentum which does not have a classical counterpart without having any coordinate representation.

We would start from the rotation of vectors in three dimensional Euclidian space and from analogy construct the quantum mechanical rotation operator that operates on a system ket to rotate it in its ket space. The quantum mechanical angular momentum can then be identified as generator of this rotation in the ket space.

7.2 Rotation in Three Dimension

We note that whereas rotations about the same axis commute but unlike translations, rotations about different axes do *not*.

We recapitulate the rotation of a vector $\mathbf{v}$ in three dimensions having cartesian components v_x, v_y, v_z. After rotation the new components v'_x, v'_y, v'_z which are given as a matrix equation:

$$\begin{pmatrix} v'_x \\ v'_y \\ v'_z \end{pmatrix} = \hat{R} \begin{pmatrix} v_x \\ v_y \\ v_z \end{pmatrix}, \tag{7.1}$$

where $\hat{R}$ is a 3×3 real orthogonal matrix which operating on a column matrix comprising of the old components v_x, v_y, v_z transforms it to the new column of elements v'_x, v'_y, v'_z. Since the length of the vector does not change, we have

$$
\begin{aligned}
v_x^2 + v_y^2 + v_z^2 &= v_x'^2 + v_y'^2 + v_z'^2, \\
\text{and}\quad \hat{R}\hat{R}^T &= \hat{R}^T\hat{R} = \hat{I}.
\end{aligned}
\tag{7.2}
$$

Specifically we consider a rotation of the vector **v** about z-axis by an angle ϕ. We will follow the convention of active rotation when the vector (or analogously the system ket) is rotated, unlike passive rotation when the coordinate axes are rotated in opposite direction. Associating the motion of a right handed screw, a positive rotation about z-axis means that the screw is advancing in the positive z-direction. It can then easily be shown that

$$
\begin{pmatrix} v'_x \\ v'_y \\ v'_z \end{pmatrix} = \begin{pmatrix} \cos\phi & -\sin\phi & 0 \\ \sin\phi & \cos\phi & 0 \\ 0 & 0 & 1 \end{pmatrix} \begin{pmatrix} v_x \\ v_y \\ v_z \end{pmatrix}.
\tag{7.3}
$$

For an infinitesimal rotation $\delta\phi$ about z-axis, the matrix $\hat{R}_z(\delta\phi)$ is given by

$$
\hat{R}_z(\delta\phi) = \begin{pmatrix} 1 - \frac{(\delta\phi)^2}{2} & -\delta\phi & 0 \\ +\delta\phi & 1 - \frac{(\delta\phi)^2}{2} & 0 \\ 0 & 0 & 1 \end{pmatrix}.
\tag{7.4}
$$

Similarly

$$
\hat{R}_x(\delta\phi) = \begin{pmatrix} 1 & 0 & 0 \\ 0 & 1 - \frac{(\delta\phi)^2}{2} & -\delta\phi \\ 0 & +\delta\phi & 1 - \frac{(\delta\phi)^2}{2} \end{pmatrix}.
\tag{7.5}
$$

and

$$
\hat{R}_y(\delta\phi) = \begin{pmatrix} 1 - \frac{(\delta\phi)^2}{2} & 0 & +\delta\phi \\ 0 & 1 & 0 \\ -\delta\phi & 0 & 1 - \frac{(\delta\phi)^2}{2} \end{pmatrix}.
\tag{7.6}
$$

Problem 7.1 Calculate the product matrices $\hat{R}_x(\delta\phi)\,\hat{R}_y(\delta\phi)$ and $\hat{R}_y(\delta\phi)\,\hat{R}_x(\delta\phi)$ and check the following:

$$
\hat{R}_x(\delta\phi)\,\hat{R}_y(\delta\phi) - \hat{R}_y(\delta\phi)\,\hat{R}_x(\delta\phi) = \begin{pmatrix} 0 & -(\delta\phi)^2 & 0 \\ (\delta\phi)^2 & 0 & 0 \\ 0 & 0 & 0 \end{pmatrix}.
\tag{7.7}
$$

Thus the matrices $\hat{R}_x(\delta\phi)$ and $\hat{R}_y(\delta\phi)$ commute with each other, if terms of the order of $(\delta\phi)^2$ or of higher order are neglected. Now since with terms up to order $(\delta\phi)^2$

$$\hat{R}_z\left((\delta\phi)^2\right) = \begin{pmatrix} 1 & -(\delta\phi)^2 & 0 \\ +(\delta\phi)^2 & 1 & 0 \\ 0 & 0 & 1 \end{pmatrix}. \tag{7.8}$$

we can write Eq. (7.7) as

$$\hat{R}_x(\delta\phi)\,\hat{R}_y(\delta\phi) - \hat{R}_y(\delta\phi)\,\hat{R}_x(\delta\phi) = \hat{R}_z\left((\delta\phi)^2\right) - \hat{I} \tag{7.9}$$

7.3 Rotation of System Kets

We shall now consider the rotation of a physical system in quantum mechanics. As the physical system is affected by rotation, the ket vector representing the system will transform accordingly. We introduce an operator $\mathcal{D}(R)$ corresponding to every physical rotation $\hat{R}$. Just as $\hat{R}$ operates on a column vector to transform it to a new column as in Eq. (7.1), so does $\mathcal{D}(R)$ on a ket vector $|\alpha\rangle$ to transform it to another ket $|\alpha\rangle_R$:

$$|\alpha\rangle_R = \mathcal{D}(R)|\alpha\rangle. \tag{7.10}$$

To obtain the explicit form of $\mathcal{D}(R)$ we examine the properties of infinitesimal rotation $\delta\phi$ in ket space. Just as in the case of infinitesimal translation dx', the unitary operator $\hat{\mathcal{T}}(d\mathbf{x}') = \hat{I} - i\hat{\mathbf{k}}\cdot d\mathbf{x}'$ was invoked in Eq. (3.25), we write

$$\mathcal{D}(R) = \hat{I} - i\hat{\mathbf{K}}\cdot\delta\boldsymbol{\phi}. \tag{7.11}$$

$\mathcal{D}(R)$ should be unitary, so that the normalization of $|\alpha\rangle$ remains unchanged. Since terms in Eq. (7.11) should be dimensionless, we define the angular momentum operator $\hat{\mathbf{J}}$ such that

$$\hat{\mathbf{K}} = \frac{\hat{\mathbf{J}}}{\hbar}. \tag{7.12}$$

$\hat{J}_n$ is a vector operator corresponding to an infinitesimal rotation $\delta\phi$ about the n-th axis. Thus

$$\mathcal{D}(\hat{\mathbf{n}}\delta\phi) = \hat{I} - i\frac{\hat{\mathbf{J}}\cdot\hat{\mathbf{n}}}{\hbar}\delta\phi \tag{7.13}$$

for a rotation $\delta\phi$ about the unit vector $\hat{\mathbf{n}}$.

A finite rotation can thus be built up by compounding successive infinitesimal rotations about the same axis. Thus

$$
\mathcal{D}_z \left(\phi \right) \;\; = \;\; \lim_{N \to \infty} \left[\hat{I} - \frac{i \hat{J}_z}{\hbar} \left(\frac{\phi}{N} \right) \right]^N \tag{7.14}
$$

$$
= \;\; \exp \left[- \frac{i \hat{J}_z \phi}{\hbar} \right] \tag{7.15}
$$

The exponential form of the operator is equivalent to an infinite series

$$
\mathcal{D}_z \left(\phi \right) \;\; = \;\; \exp \left[- \frac{i \hat{J}_z \phi}{\hbar} \right]
$$

$$
= \;\; \hat{I} + \frac{1}{1!} \left(\frac{-i \hat{J}_z \phi}{\hbar} \right)^1 + \frac{1}{2!} \left(\frac{-i \hat{J}_z \phi}{\hbar} \right)^2 + \frac{1}{3!} \left(\frac{-i \hat{J}_z \phi}{\hbar} \right)^3 + \; \cdots \; . \tag{7.16}
$$

Thus for every rotation $\hat{R}$ in the Euclidian space, we can identify a rotation operator $\mathcal{D} \left(R \right)$ analogously in the relevant ket space. We extend this correspondence of ket rotation with the vector rotations in Section 7.2 by postulating the same group properties for $\mathcal{D} \left(R \right)$ as for $\hat{R}$. Thus we have:

(i). Identity:

$$
\hat{R} \cdot \hat{I} = \hat{R} \;\; \Longrightarrow \;\; \mathcal{D} \left(R \right) \cdot \hat{I} = \mathcal{D} \left(R \right) . \tag{7.17}
$$

(ii). Closure:

$$
\hat{R}_1 \hat{R}_2 = \hat{R}_3 \;\; \Longrightarrow \;\; \mathcal{D} \left(R_1 \right) \mathcal{D} \left(R_2 \right) = \mathcal{D} \left(R_3 \right) . \tag{7.18}
$$

(iii). Inverse:

$$
\hat{R} \hat{R}^{-1} = \hat{I} \;\; \Longrightarrow \;\; \mathcal{D} \left(R \right) \mathcal{D}^{-1} \left(R \right) = \hat{I}, \tag{7.19}
$$

$$
\text{and } \hat{R}^{-1} \hat{R} = \hat{I} \;\; \Longrightarrow \;\; \mathcal{D}^{-1} \left(R \right) \mathcal{D} \left(R \right) = \hat{I}. \tag{7.20}
$$

(iv). Associativity:

$$
\hat{R}_1 \left(\hat{R}_2 \hat{R}_3 \right) = \left(\hat{R}_1 \hat{R}_2 \right) \hat{R}_3 = \hat{R}_1 \hat{R}_2 \hat{R}_3 \;\; \Longrightarrow \;\; \mathcal{D} \left(R_1 \right) \left[\mathcal{D} \left(R_2 \right) \mathcal{D} \left(R_3 \right) \right] =
$$

$$
\left[\mathcal{D} \left(R_1 \right) \mathcal{D} \left(R_2 \right) \right] \mathcal{D} \left(R_3 \right) \;\; = \;\; \mathcal{D} \left(R_1 \right) \mathcal{D} \left(R_2 \right) \mathcal{D} \left(R_3 \right) . \tag{7.21}
$$

Then extending this analogy to Eq. (7.9) and making use of Eq. (7.13) we obtain up to terms quadratic in $\delta\phi$

$$\left[\hat{I} - i\frac{\hat{J}_x}{\hbar}(\delta\phi) - \frac{\hat{J}_x^2}{2\hbar^2}(\delta\phi)^2 + \cdots\right]\left[\hat{I} - i\frac{\hat{J}_y}{\hbar}(\delta\phi) - \frac{\hat{J}_y^2}{2\hbar^2}(\delta\phi)^2 + \cdots\right]$$

$$- \left[\hat{I} - i\frac{\hat{J}_y}{\hbar}(\delta\phi) - \frac{\hat{J}_y^2}{2\hbar^2}(\delta\phi)^2 + \cdots\right]\left[\hat{I} - i\frac{\hat{J}_x}{\hbar}(\delta\phi) - \frac{\hat{J}_x^2}{2\hbar^2}(\delta\phi)^2 + \cdots\right]$$

$$= \hat{I} - i\frac{\hat{J}_z}{\hbar^2}(\delta\phi)^2 - \hat{I}. \tag{7.22}$$

Equating terms of the order of $(\delta\phi)$ and $(\delta\phi)^2$ separately, we get

$$\left[\hat{J}_x, \hat{J}_y\right] = i\hbar\hat{J}_z, \tag{7.23}$$

as the terms of the order of $(\delta\phi)$ cancel out. In general we have

$$\left[\hat{J}_i, \hat{J}_j\right] = i\hbar\epsilon_{ijk}\hat{J}_k, \tag{7.24}$$

where $\hat{J}_i$, $\hat{J}_j$, $\hat{J}_k$ are the Cartesian components of $\hat{\mathbf{J}}$ and ϵ_{ijk} is the permutation symbol defined by

$$\epsilon_{ijk} = \begin{cases} +1 & \text{if } i,j,k \text{ is an even permutation of } 1,2,3 \\ -1 & \text{if } i,j,k \text{ is an odd permutation of } 1,2,3 \\ 0 & \text{if } \qquad \text{any two indices are equal} \end{cases}. \tag{7.25}$$

7.4   Eigenvalue and Eigenvectors of Angular Momentum

We have defined the angular momentum $\hat{J}_i$ as the generator of infinitesimal rotation about the i-th axis. We now define $\hat{J}^2$ as

$$\hat{J}^2 = \hat{J}_x\hat{J}_x + \hat{J}_y\hat{J}_y + \hat{J}_z\hat{J}_z = \hat{J}_x^2 + \hat{J}_y^2 + \hat{J}_z^2. \tag{7.26}$$

Problem 7.2 Use Eq. (7.24) and show that

$$\left[\hat{J}^2, \hat{J}_k\right] = 0, \quad k = 1,\ 2,\ 3. \tag{7.27}$$

Since $\hat{J}^2$ commutes with each of $\hat{J}_x$, $\hat{J}_y$, $\hat{J}_z$ but $\hat{J}_x$, $\hat{J}_y$, $\hat{J}_z$ do not commute with each other, we can choose $\hat{J}^2$ and $\hat{J}_z$ and find their common eigenkets

$$\hat{J}^2|a,b\rangle \ = \ a|a,b\rangle, \tag{7.28}$$
$$\hat{J}_z|a,b\rangle \ = \ b|a,b\rangle. \tag{7.29}$$

We now define the *ladder operators*

$$\hat{J}_\pm \ = \ \hat{J}_x \pm i\hat{J}_y. \tag{7.30}$$

Evidently $\hat{J}_\pm$ are not Hermitian.

Problem 7.3 Show that

$$\left[\hat{J}_+, \hat{J}_-\right] \ = \ 2\hbar\hat{J}_z, \tag{7.31}$$
$$\hat{J}_\pm\hat{J}_\mp \ = \ \hat{J}^2 - \hat{J}_z^2 \pm \hbar\hat{J}_z, \tag{7.32}$$
$$\text{and} \ \left[\hat{J}_z, \hat{J}_\pm\right] \ = \ \pm\hbar\hat{J}_\pm. \tag{7.33}$$

Now we have from Eq. (7.33)

$$\begin{aligned}
\hat{J}_z\hat{J}_+|a,b\rangle \ &= \ \hat{J}_+\hat{J}_z|a,b\rangle + \hbar\hat{J}_+|a,b\rangle \\
&= \ b\hat{J}_+|a,b\rangle + \hbar\hat{J}_+|a,b\rangle \\
&= \ (b+\hbar)\,\hat{J}_+|a,b\rangle. \tag{7.34}
\end{aligned}$$

Similarly

$$\hat{J}_z\hat{J}_-|a,b\rangle \ = \ (b-\hbar)\,\hat{J}_-|a,b\rangle. \tag{7.35}$$

Thus $\hat{J}_\pm|a,b\rangle$ are the eigenkets of $\hat{J}_z$ belonging to eigenvalues $(b \pm \hbar)$. In other words $\hat{J}_+$ ($\hat{J}_-$) raises (lowers) the eigenvalue b by $\hbar$, without changing the eigenvalue a of $\hat{J}^2$. Since $\hat{J}^2$ commutes with $\hat{J}_x$ and $\hat{J}_y$ we can write

$$\begin{aligned}
\hat{J}^2\left[\hat{J}_\pm|a,b\rangle\right] \ &= \ \hat{J}_\pm\left[\hat{J}^2|a,b\rangle\right] \\
&= \ a\left[\hat{J}_\pm|a,b\rangle\right]. \tag{7.36}
\end{aligned}$$

Thus $\hat{J}_\pm|a,b\rangle$ is the simultaneous eigenket of $\hat{J}^2$ and $\hat{J}_z$ with eigenvalues of a and $b \pm \hbar$ respectively. We thus have

$$\hat{J}_\pm|a,b\rangle \ \propto \ |a,b\pm\hbar\rangle \ = \ N_\pm|a,b\pm\hbar\rangle. \tag{7.37}$$

$N_\pm$ is the normalization constant.

Since $\hat{J}^2 = \hat{J}_x^2 + \hat{J}_y^2 + \hat{J}_z^2$ and because the expectation values of the squares of Hermitian operators must be positive or zero, we have

$$\langle \hat{J}^2 \rangle = \langle \hat{J}_x^2 \rangle + \langle \hat{J}_y^2 \rangle + \langle \hat{J}_z^2 \rangle,$$

$$\text{and } \langle \hat{J}^2 \rangle \geq \langle \hat{J}_z^2 \rangle,$$

$$\text{so} \quad a \geq b^2. \tag{7.38}$$

If we operate $\hat{J}_+$ on any eigenket $|a, b\rangle$ of $\hat{J}^2$ and $\hat{J}_z$ successively n times then we shall get another eigenket of $\hat{J}^2$ and $\hat{J}_z$ with the eigenvalue of $\hat{J}_z$ increased by $n\hbar$ but the eigenvalue of $\hat{J}^2$ remaining unchanged. But because of the inequality Eq. (7.38) there exists an upper limit to the eigenvalue b of $\hat{J}_z$, designated as $b_{\max}$ such that

$$\hat{J}_+ |a, b_{\max}\rangle = 0, \tag{7.39}$$
$$\hat{J}_- \hat{J}_+ |a, b_{\max}\rangle = 0. \tag{7.40}$$

Since

$$\hat{J}_- \hat{J}_+ = \hat{J}_x^2 + \hat{J}_y^2 - i\left(\hat{J}_y \hat{J}_x - \hat{J}_x \hat{J}_y\right) = \hat{J}^2 - \hat{J}_z^2 - \hbar \hat{J}_z, \tag{7.41}$$

we have

$$\left(\hat{J}^2 - \hat{J}_z^2 - \hbar \hat{J}_z\right) |a, b_{\max}\rangle = 0,$$

$$a - b_{\max}^2 - \hbar b_{\max} = 0. \tag{7.42}$$

Similarly from the same argument there is a lower limit $b_{\min}$ of b such that

$$\hat{J}_- |a, b_{\min}\rangle = 0,$$
$$\hat{J}_+ \hat{J}_- |a, b_{\min}\rangle = 0. \tag{7.43}$$

and since

$$\hat{J}_+ \hat{J}_- = \hat{J}^2 - \hat{J}_z^2 + \hbar \hat{J}_z, \tag{7.44}$$

we get from Eq. (7.43)

$$a - b_{\min}^2 + \hbar b_{\min} = 0, \tag{7.45}$$

and from Eq. (7.42) and Eq. (7.45) we have

$$(b_{\max} + b_{\min})(b_{\max} - b_{\min} + \hbar) = 0. \tag{7.46}$$

The equality Eq. (7.46) is satisfied for

$$b_{\max} = -b_{\min}; \tag{7.47}$$

the other solution is not physically admissible.

Thus

$$-b_{\max} \leq b \leq b_{\max}. \tag{7.48}$$

It is then possible to attain the eigenstate $|a, b_{\max}\rangle$ starting from $|a, -b_{\max}\rangle$ by successively applying $\hat{J}_+$ on $|a, -b_{\max}\rangle$ say n times and we have

$$
\begin{aligned}
b_{\max} &= -b_{\max} + n\hbar, \quad \text{or} \\
b_{\max} &= \frac{n}{2}\hbar, \quad n = 1, 2, \cdots .
\end{aligned}
\tag{7.49}
$$

Putting $j = n/2$,

$$b_{\max} = j\hbar, \quad j = \frac{1}{2}, 1, \frac{3}{2}, 2, \cdots , \quad \text{integers or half odd integers.} \tag{7.50}$$

From Eq. (7.42) we get

$$a = b_{\max} (b_{\max} + \hbar) = j(j+1)\hbar^2. \tag{7.51}$$

Since $-b_{\max} \leq b \leq b_{\max}$ we can define m such that

$$b = m\hbar \tag{7.52}$$

It then follows that the allowed values of m for a given value of j are

$$m = -j, -j+1, \cdots , j-1, j, \tag{7.53}$$

which are $2j + 1$ in number. In other words for a given value of *total angular momentum quantum number* j there are $2j + 1$ m-states for $\hat{J}_z$. The eigenvalue equations for $\hat{J}^2$ and $\hat{J}_z$ are

$$
\begin{aligned}
\hat{J}^2|j,m\rangle &= j(j+1)\hbar^2|j,m\rangle, \\
j &= \frac{1}{2}, 1, \frac{3}{2}, 2, \cdots ,
\end{aligned}
\tag{7.54}
$$

$$
\begin{aligned}
\hat{J}_z|j,m\rangle &= m\hbar|j,m\rangle, \\
m &= -j, -j+1, \cdots , j+1, j .
\end{aligned}
\tag{7.55}
$$

7.5 Matrix Representation of Angular Momentum Operator

The simultaneous eigenkets $|j, m\rangle$ of $\hat{J}^2$ and $\hat{J}_z$ form a orthonormal and complete set of states and can be used as basis vectors to obtain the matrix representation of angular momentum operators. Obviously $\hat{J}^2$ and $\hat{J}_z$ are diagonal in this representation:

$$\langle j', m'|\hat{J}^2|j, m\rangle = j(j+1)\hbar^2 \delta_{j,j'}\delta_{m,m'}, \tag{7.56}$$

$$\langle j', m'|\hat{J}_z|j, m\rangle = m\hbar \delta_{j,j'}\delta_{m,m'}, \tag{7.57}$$

To obtain matrices for $\hat{J}_x$ and $\hat{J}_y$ we use

$$\hat{J}_\pm = \hat{J}_x \pm i\hat{J}_y.$$

We have from Eq. (7.37)

$$\hat{J}_\pm|j, m\rangle = N_\pm|j, m \pm 1\rangle, \tag{7.58}$$

where $N_\pm$ are the normalization constants and we have

$$|N_\pm|^2 = \langle j, m|\hat{J}_\pm^\dagger \hat{J}_\pm|j, m\rangle \tag{7.59}$$

$$= \langle j, m|\hat{J}_\mp \hat{J}_\pm|j, m\rangle \tag{7.60}$$

$$= \langle j, m|\left(\hat{J}^2 - \hat{J}_z^2 \mp \hbar\hat{J}_z\right)|j, m\rangle \tag{7.61}$$

$$= [j(j+1) - m(m \pm 1)]\hbar^2. \tag{7.62}$$

$$N_\pm = \sqrt{j(j+1) - m(m \pm 1)}\hbar. \tag{7.63}$$

Thus

$$\langle j', m'|\hat{J}_\pm|j, m\rangle = \sqrt{j(j+1) - m(m \pm 1)}\hbar \delta_{j',j}\delta_{m',m\pm1}. \tag{7.64}$$

In other words

$$\left(\hat{J}_+\right)_{j',m';j.m} = \sqrt{j(j+1) - m(m+1)}\hbar \delta_{j',j}\delta_{m',m+1}, \tag{7.65}$$

$$\left(\hat{J}_-\right)_{j',m';j.m} = \sqrt{j(j+1) - m(m-1)}\hbar \delta_{j',j}\delta_{m',m-1}. \tag{7.66}$$

And we have

$$\hat{J}_x = \frac{1}{2}\left[\hat{J}_+ + \hat{J}_-\right], \tag{7.67}$$

$$\hat{J}_y = \frac{1}{2i}\left[\hat{J}_+ - \hat{J}_-\right]. \tag{7.68}$$

We can now study the matrix elements of the rotation operator $\mathcal{D}(R)$. For a rotation $\hat{R}$ specified by $\hat{\mathbf{n}}$ and ϕ, the matrix elements can be defined by

$$\mathcal{D}^j_{m',m}(R) = \langle j,m'|\exp\left[-i\frac{\hat{\mathbf{J}}\cdot\hat{\mathbf{n}}}{\hbar}\phi\right]|j,m\rangle. \tag{7.69}$$

These are sometimes called *Wigner functions* after E.P.Wigner whose contributions to the group theoretical properties of rotations in quantum mechanics are pioneering. We notice that the matrix elements in Eq. (7.69) is diagonal in the quantum number j. This is because $\mathcal{D}(R)|j,m\rangle$ is still an eigenket of $\hat{J}^2$ with the eigenvalue $j(j+1)\hbar^2$:

$$\hat{J}^2[\mathcal{D}(R)|j,m\rangle] = \mathcal{D}(R)\hat{J}^2|j,m\rangle = j(j+1)\hbar^2[\mathcal{D}(R)|j,m\rangle]. \tag{7.70}$$

This is so because $\hat{J}^2$ commutes all $\hat{J}_k$ (and therefore with any function of $\hat{J}_k$). In other words rotations cannot change the total angular momentum quantum number j, only the projection quantum numbers change.

The $(2j+1)\times(2j+1)$ matrix $\mathcal{D}^j_{m',m}(R)$ is referred to as the $(2j+1)$ dimensional irreducible reprentation of the rotation operator R_j.

It is evident that matrices of $\hat{J}_x$, $\hat{J}_y$, $\hat{J}_z$, $\hat{J}_+$, $\hat{J}_-$ and $\hat{J}^2$ are all diaginal in the quantum number j. We can thus construct an infinite number of representations for these matrices corresponding to the values of $j = \frac{1}{2}, 1, \frac{3}{2}, \cdots$ and having $(2j+1)$ columns and rows labelled by the values of m and m' respectively. All these representations can be taken together to form one single representation of infinite rank with finite dimensional blocks for each j at the diagonal position.

7.6   Orbital Angular Momentum

In classical mechanics a particle with linear momentum $\mathbf{p}$ having a position coordinate $\mathbf{r}$ has an *angular momentum* $\mathbf{L}$ about the origin

$$\mathbf{L} = \mathbf{r}\times\mathbf{p}, \tag{7.71}$$

with Cartesian components

$$L_x = yp_z - zp_y, \tag{7.72}$$
$$L_y = zp_x - xp_z, \tag{7.73}$$
$$L_x = xp_y - yp_x. \tag{7.74}$$

Since the quantum angular momentum is defined by the commutation relations Eq. (7.24), it is pertinent to verify that Eq. (7.72), Eq. (7.73) and Eq. (7.74) satisfy the same.

Problem 7.4 Use the fundamental commutation relations between $\hat{p}_i$ and $\hat{x}_j$ in Eq. (3.48) to show

(i). $\quad \left[\hat{L}_i, \hat{L}_j \right] = i\hbar\epsilon_{ijk}\hat{L}_k.$ $\hfill (7.75)$

(ii). From the relation $\hat{L}^2 = \hat{L}_x^2 + \hat{L}_y^2 + \hat{L}_z^2$, show that

$$\left[\hat{L}^2, \hat{L}_i \right] = 0. \text{ where } i = 1, 2, 3 \text{ (i.e. } x, y, z).$$ $\hfill (7.76)$

EIGENVALUES AND EIGENFUNCTIONS OF $\hat{L}_z$ and $\hat{L}^2$.

We have

$$\hat{L}_z = \hat{x}\hat{p}_y - \hat{y}\hat{p}_x.$$ $\hfill (7.77)$

We would work out in coordinate representation for $\hat{p}_x$ and $\hat{p}_y$. We thus have

$$\hat{L}_z = -i\hbar \left(x\frac{\partial}{\partial y} - y\frac{\partial}{\partial x} \right).$$ $\hfill (7.78)$

Since

$$x = r\sin\theta\cos\phi,$$
$$y = r\sin\theta\sin\phi,$$
$$z = r\cos\theta,$$

we may transform to spherical polar coordinates.

Problem 7.5 Use the inverse transformations

$$r^2 = x^2 + y^2 + z^2,$$
$$\cos\theta = \frac{y}{r} = \frac{y}{\sqrt{x^2 + y^2 + z^2}},$$
$$\tan\phi = \frac{y}{x}$$

and obtain the following relations

(i). $\quad \hat{L}_z = i\hbar\frac{\partial}{\partial\phi},$ $\hfill (7.79)$

(ii). $\quad \hat{L}_x = i\hbar \left[\sin\phi\frac{\partial}{\partial\theta} + \cot\theta\cos\phi\frac{\partial}{\partial\phi} \right],$ $\hfill (7.80)$

(iii). $\hat{L}_y = i\hbar \left[-\cos\phi \dfrac{\partial}{\partial\theta} + \cot\theta \sin\phi \dfrac{\partial}{\partial\phi} \right],$ (7.81)

(iv). and hence

$$\hat{L}^2 = -\hbar^2 \left[\frac{1}{\sin\theta} \frac{\partial}{\partial\theta} \left(\sin\theta \frac{\partial}{\partial\theta} \right) + \frac{1}{\sin^2\theta} \left(\frac{\partial}{\partial\phi} \right)^2 \right].$$ (7.82)

[You may use the following relationships.]

$$\frac{\partial r}{\partial z} = \cos\theta, \qquad \frac{\partial r}{\partial y} = \sin\theta \sin\phi, \qquad \frac{\partial r}{\partial x} = \sin\theta \cos\phi,$$

$$\frac{\partial\theta}{\partial z} = -\frac{\sin\theta}{r}, \qquad \frac{\partial\theta}{\partial y} = \frac{\cos\theta \sin\phi}{r}, \qquad \frac{\partial\theta}{\partial x} = \frac{\cos\theta \cos\phi}{r},$$

$$\frac{\partial\phi}{\partial z} = 0, \qquad \frac{\partial\phi}{\partial y} = \frac{\cos\phi}{r\sin\theta}, \qquad \frac{\partial\phi}{\partial x} = -\frac{\sin\phi}{r\sin\theta}.$$

Since $\hat{L}_z = i\hbar\frac{\partial}{\partial\phi}$, the eigenvalue equation becomes

$$i\hbar\frac{\partial}{\partial\phi}\Phi(\phi) = l_z\Phi(\phi),$$ (7.83)

where the azimuthal angle ϕ lies between 0 and 2π, i.e. $0 \le \phi \le 2\pi$ and l_z is the eigenvalue.

The solution of Eq. (7.83) is

$$\Phi(\phi) = N \cdot \exp\left(\frac{il_z\phi}{\hbar} \right)$$ (7.84)

Since Φ is to be a single valued function of ϕ and $\Phi(\phi) = \Phi(\phi + 2\pi)$ i.e.

$$\exp\left(\frac{il_z\phi}{\hbar} \right) = \exp\left(\frac{il_z(\phi + 2\pi)}{\hbar} \right),$$

so $l_z = m\hbar, \quad m = 0, \pm 1, \pm 2, \cdots,$

$$\Phi_m(\phi) = N\exp(im\phi),$$ (7.85)

and $|N|^2 \displaystyle\int_0^{2\pi} \Phi_m^*(\phi)\,\Phi_m(\phi)\,d\phi = 1,$ (7.86)

will give

$$|N|^2 = \frac{1}{2\pi}.$$

$$\therefore \quad \Phi_m(\phi) = \frac{1}{\sqrt{2\pi}}\exp(im\phi), \text{ with } m = 0, \pm 1, \pm 2, \cdots.$$ (7.87)

The eigenvalue equation for $\hat{L}^2$ is given by

$$-\hbar^2 \left[\frac{1}{\sin\theta} \frac{\partial}{\partial\theta} \left(\sin\theta \frac{\partial}{\partial\theta} \right) + \frac{1}{\sin^2\theta} \left(\frac{\partial}{\partial\phi} \right)^2 + \frac{L^2}{\hbar^2} \right] \Psi\left(\theta,\phi\right) = 0, \tag{7.88}$$

where L^2 is the eigenvalue of $\hat{L}^2$.

We compare this with the one satisfied by the *Spherical Harmonics* $Y_{l,m}\left(\theta,\phi\right)$:

$$\left[\frac{1}{\sin\theta} \frac{\partial}{\partial\theta} \left(\sin\theta \frac{\partial}{\partial\theta} \right) + \frac{1}{\sin^2\theta} \left(\frac{\partial}{\partial\phi} \right)^2 + l\left(l+1\right) \right] Y_{l,m}\left(\theta,\phi\right) = 0, \tag{7.89}$$

where $l = 0, 1, 2, \cdots$; and $|m| = 0, 1, 2, \cdots, l$.

Thus the two equations become identical if the eigenvalue L^2 is set equal to

$$L^2 = l\left(l+1\right)\hbar^2. \tag{7.90}$$

The explicit dependence of the Spherical harmonics on θ and ϕ for positive values of m is given as

$$Y_{l,m}\left(\theta,\phi\right) = \Theta_{l,m}\left(\theta\right) \frac{1}{\sqrt{2\pi}} \exp\left(im\phi\right). \tag{7.91}$$

The real valued function $\Theta_{l,m}$ can be expressed as

$$\Theta_{l,m}\left(\theta,\phi\right) = \left(-1\right)^m \sqrt{\frac{\left(2l+1\right)\left(l-m\right)!}{2\left(l+m\right)!}} \left(\sin\theta\right)^m \left(\frac{d}{d\cos\theta} \right)^m P_l\left(\cos\theta\right) \tag{7.92}$$

$$\text{where } P_l\left(\cos\theta\right) = \frac{1}{l!} \left(\frac{d}{d\cos\theta} \right)^l \left(\cos^2\theta - 1\right)^l \tag{7.93}$$

are called the *Legendre polynomials*.

Spherical harmonics for the negative $m = -l, -\left(l-1\right), \cdots, -2, -1$ are defined by the condition

$$Y_{l,-m}\left(\theta,\phi\right) = \left(-1\right)^m Y_{l,m}^*\left(\theta,\phi\right). \tag{7.94}$$

This particular phase of the Spherical Harmonics is called Condon-Shortley phase. With this phase the expression of the orthonormalized Spherical Harmonics is

$$Y_{l,m}\left(\theta,\phi\right) = \left(-1\right)^{\frac{m}{2}+\frac{|m|}{2}} \sqrt{\left[\frac{2l+1}{4\pi} \frac{\left(l-|m|\right)!}{\left(l+|m|\right)!} \right]} \frac{e^{im\phi} \sin^{|m|}\theta}{2^l l!} \times$$

$$\left(\frac{d}{d\cos\theta} \right)^{l+|m|} \left(\cos^2\theta - 1\right)^l. \tag{7.95}$$

Since $Y_{l,m}(\theta, \phi)$ is the simultaneous eigenfunction of $\hat{L}_z = -i\hbar\frac{\partial}{\partial\phi}$ and $\hat{L}^2$ and it is evident that

$$\hat{L}_z\, Y_{l,m}(\theta, \phi) \;=\; m\hbar\, Y_{l,m}(\theta, \phi)\,, \text{ with } m = -l, -l+1, \cdots, l-1, l,$$
$$\tag{7.96}$$

$$\hat{L}^2\, Y_{l,m}(\theta, \phi) \;=\; l(l+1)\,\hbar^2\, Y_{l,m}(\theta, \phi)\,, \text{ with } l = 0, 1, 2, \cdots, \tag{7.97}$$

The Spherical harmonics are orthogonal and normalized eigenfunctions of the commuting observables $\hat{L}_z$ and $\hat{L}^2$ with orthonormality conditions

$$\int_{\theta=0}^{\theta=\pi} \int_{\phi=0}^{\phi=2\pi} Y_{l,m}^*(\theta, \phi)\, Y_{l',m'}(\theta, \phi) \sin\theta\; d\theta d\phi \;=\; \delta_{l,l'}\delta_{m.m'}. \tag{7.98}$$

The first few spherical harmonics are given in Table (7.1)

Table 7.1: Tables of Spherical Harmonics for $l = 0,\ 1,\ 2$.

$$Y_{0,0} \;=\; \sqrt{\frac{1}{4\pi}}; \tag{7.99}$$

$$Y_{1,\pm1} \;=\; \mp\sqrt{\frac{3}{8\pi}}\sin\theta \exp(\pm i\phi)\,; \tag{7.100}$$

$$Y_{1,0} \;=\; \sqrt{\frac{3}{4\pi}}\cos\theta; \tag{7.101}$$

$$Y_{2,\pm2} \;=\; \frac{1}{4}\sqrt{\frac{15}{2\pi}}\sin^2\theta \exp(\pm 2i\phi)\,; \tag{7.102}$$

$$Y_{2,\pm1} \;=\; \mp\sqrt{\frac{15}{8\pi}}\sin\theta\cos\theta \exp(\pm i\phi)\,; \tag{7.103}$$

$$Y_{2,0} \;=\; \sqrt{\frac{5}{4\pi}}\left(\frac{3}{2}\cos^2\theta - \frac{1}{2}\right). \tag{7.104}$$

Chapter 8

Spin Angular Momentum

Early evidences show that elementary particles like electrons possess an intrinsic degree of freedom which is akin to angular momentum but has no classical description like orbital angular momentum. This is termed *spin* and is denoted by *s*. O. Stern and W. Gerlach demonstrated in 1922 that electron has a spin 1/2. We shall consider this landmark experiment in the next section.

8.1 The Stern Gerlach Experiment

The Stern Gerlach experiment originally conceived by O. Stern and carried out in collaboration with W. Gerlach, illustrates in a striking manner the necessity for a radical departure from the concepts of classical mechanics and firmly established the quantum nature of spin. In a sense the two-state spin $\frac{1}{2}$ system is the least classical, most quantum system and is often cited for its simplicity and clarity.

A schematic diagram is given in Fig. (8.1).

A beam of paramagnetic silver atom Ag^{47} is produced in an oven and collimated by allowing the beam to go through a collimator. The beam is then subjected to an inhomogeneous magnetic field by a properly shaped magnetic pole pieces. After passing through the magnetic field the beam goes to the detector.

A silver atom is composed of 47 electrons and the nucleus. 46 of these electrons form a spherically symmetric charge distribution having no angular momentum. The 47-th electron is a $5S$ electron according to spectroscopic notation having orbital angular momentum $l = 0$. Thus ignoring the nuclear spin, the atom has the angular momentum $l = 0$. For a magnetic field in the z-direction we would expect classically to see a continuous distribution on the detector screen about the undeflected direction $z = 0$. In case of a $5S$ electron, no splitting should occur and there should be one spot on the screen as is shown in Fig. (8.1) as 'Expected from Classical Physics'. In the experiment itself, the beam was split into two distinct components, described

The Stern Gerlach Arrangement

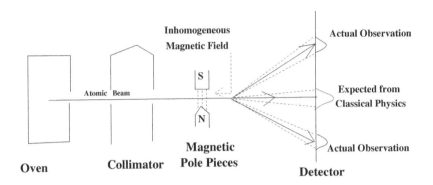

Figure 8.1: Schematic arrangement of Stern Gerlach Experiment.

as 'Actual Observation'. To explain this, it was postulated that the electron, in addition to the orbital angular momentum, possesses an intrinsic spin angular momentum, which can assume only the values $+\hbar/2$ and $-\hbar/2$ about an arbitrarily chosen direction. The 47 electrons are attached to the nucleus which is very heavy (about 2×10^5 times heavier than electron) and thus the atom as whole possesses in Gaussian units a magnetic moment $\boldsymbol{\mu}_S = g_S \frac{e}{2mc} \boldsymbol{S}$. Here e = electronic charge in e.s.u., m = electronic mass, c = the speed of light in vacuum, and for electron $g_S = 2$. The force on the electron passing through an inhomogeneous magnetic field B_z along the z-direction is

$$\mathbf{F} = -\boldsymbol{\nabla}\left(-\boldsymbol{\mu}\cdot\mathbf{B}\right) \approx \mu_z \frac{\partial B_z}{\partial z}\hat{\mathbf{z}}, \qquad (8.1)$$

ignoring the component of the magnetic field $\mathbf{B}$ in directions other than the z-direction.

The beam is then expected to get split according to the values of μ_z. Thus the SG apparatus "measures" the z component of μ or equivalently the z-component of spin S up to a proportionality factor.

The atoms coming out of the oven are randomly oriented. If the electrons were like classical objects, we would expect all values of μ_z to be present between $+|\mu|$ and $-|\mu|$. This would have given a continuous bundle of beams coming out of the apparatus as shown in the central position of the detector screen. Instead, we observe that the beam coming out is split in two distinct components corresponding to two possible values of the z-component of S: S_z up and S_z down which we call $S_{z\uparrow}$ and $S_{z\downarrow}$ respectively. Thus the SG experiment provides the evidence of the existence of a spin degree of freedom of electron.

8.2 Matrix Representation of Spin

Quantum mechanically we have already obtained the first non-trivial angular momentum quantum number $j = \frac{1}{2}$ in § (7.4).

We shall use the the notation $\hat{\boldsymbol{S}}$ for spin operator and $\hat{S}_z$ for its component in place of $\hat{\boldsymbol{J}}$ and $\hat{J}_z$ which are general notation for angular momentum defined by the commutation relation Eq. (7.24). Similarly we shall use s in place of j and m_s in place of m. We can work out using the formalism developed in § (7.5) the matrix forms for the various operators $\hat{\boldsymbol{S}}^2$, $\hat{S}_z$, $\hat{S}_\pm$ accordingly for $s = \frac{1}{2}$. For $s = \frac{1}{2}$, we have $m_s = +\frac{1}{2}$ and $m_s = -\frac{1}{2}$, and accordingly the two eigenkets of $\hat{\boldsymbol{S}}^2$ and $\hat{S}_z$ are $|s = \frac{1}{2}, m_s = +\frac{1}{2}\rangle$ and $|s = \frac{1}{2}, m_s = -\frac{1}{2}\rangle$. They are orthonormal column vectors.

$$|s = \frac{1}{2}, m_s = +\frac{1}{2}\rangle = \begin{pmatrix} 1 \\ 0 \end{pmatrix}, \quad \text{and} \quad |s = \frac{1}{2}, m_s = -\frac{1}{2}\rangle = \begin{pmatrix} 0 \\ 1 \end{pmatrix}. \tag{8.2}$$

Similar to Eq. (7.56) and Eq. (7.57) we now have

$$\begin{aligned} \langle s', s'_m | \hat{S}^2 | s, s_m \rangle &= s(s+1)\hbar^2 \delta_{s,s'} \delta_{m_s, m'_s}, \\ \langle s', m'_s | \hat{S}_z | s, m_s \rangle &= m_s \hbar \delta_{s,s'} \delta_{m_s, m'_s}, \end{aligned}$$

From these we can construct the diagonal matrices for $\hat{S}^2$ and $\hat{S}_z$ in this basis and obtain the matrices

$$\hat{S}^2 = \frac{3}{4}\hbar^2 \begin{pmatrix} 1 & 0 \\ 0 & 1 \end{pmatrix}, \quad \text{and} \quad \hat{S}_z = \frac{1}{2}\hbar \begin{pmatrix} 1 & 0 \\ 0 & -1 \end{pmatrix}. \tag{8.3}$$

Similarly from Eq. (7.65) Eq. (7.66) we can write

$$\hat{S}_+ = \hbar \begin{pmatrix} 0 & 1 \\ 0 & 0 \end{pmatrix}, \tag{8.4}$$

$$\hat{S}_- = \hbar \begin{pmatrix} 0 & 0 \\ 1 & 0 \end{pmatrix}, \tag{8.5}$$

It is then straightforward to obtain the matrices for

$$\hat{S}_x = \frac{1}{2}\left(\hat{S}_+ + \hat{S}_-\right) = \frac{1}{2}\hbar \begin{pmatrix} 0 & 1 \\ 1 & 0 \end{pmatrix}, \tag{8.6}$$

$$\hat{S}_y = \frac{1}{2i}\left(\hat{S}_+ - \hat{S}_-\right) = \frac{1}{2}\hbar \begin{pmatrix} 0 & -i \\ i & 0 \end{pmatrix}, \tag{8.7}$$

8.3   Finite Rotations in Spin-$\frac{1}{2}$ Space

We have obtained the eigenvectors of $\hat{S}^2$ and $\hat{S}_z$ for spin $\frac{1}{2}$ in the previous section. The spin space is of demension 2 and the eigenkets $|\frac{1}{2}, +\frac{1}{2}\rangle$ and $|\frac{1}{2}, -\frac{1}{2}\rangle$ can be taken as the basis kets. Any arbitrary ket in this space can be expressed as a superposition of these two base kets. We designate the spin up ket by $|+\rangle$ and the spin doown by $|-\rangle$. Then the completeness condition given by the identity operator according to Eq. (1.35), which in this case is

$$|+\rangle\langle+| + |-\rangle\langle-| = \hat{I}. \tag{8.8}$$

Then

$$\hat{S}_z = \hat{S}_z \left[|+\rangle\langle+| + |-\rangle\langle-| \right] \quad \text{implies}$$
$$\hat{S}_z = \frac{1}{2}\hbar \left[|+\rangle\langle+| - |-\rangle\langle-| \right]. \tag{8.9}$$

Problem 8.1 The state of a spin-$\frac{1}{2}$ particle is given by

$$|\alpha\rangle = \frac{1}{2}|+\rangle + \frac{\sqrt{3}}{2}|-\rangle, \quad \text{where } S_z|\pm\rangle = \pm\frac{\hbar}{2}|\pm\rangle.$$

(i). What is the probability that the particle has a spin $-\frac{\hbar}{2}$ in the state $|\alpha\rangle$?

(ii). Obtain the expectation value of $\hat{S}_z$ in the above state?

Problem 8.2 Show that the non-Hermitian matrices $\hat{S}+$ and $\hat{S}_-$ become

(i). $\hat{S}_+ = \hbar|+\rangle\langle-|.$ \hfill (8.10)

(ii). $\hat{S}_- = \hbar|-\rangle\langle+|.$ \hfill (8.11)

(iii). Using Eq. (8.10) and Eq. (8.11), show that

$$\hat{S}_x = \frac{1}{2}\hbar \left[|+\rangle\langle-| + |-\rangle\langle+| \right], \tag{8.12}$$

$$\hat{S}_y = \frac{1}{2i}\hbar \left[|+\rangle\langle-| - |-\rangle\langle+| \right], \tag{8.13}$$

(iv). Also show that $\hat{S}_x$ and $\hat{S}_y$ are Hermitian.

$\hat{S}_+$ and $\hat{S}_-$ are the *raising* and *lowering* operators respectively that raises and lowers the spin component by one unit of $\hbar$.

We now consider a rotation of the ket $|\alpha\rangle$ of a spin-$\frac{1}{2}$ system by a finite angle ϕ about the z-axis. Then from Eq. (7.10)

$$|\alpha\rangle_R = \mathcal{D}_z(\phi)|\alpha\rangle, \quad \text{with} \tag{8.14}$$

$$\mathcal{D}_z(\phi) = \exp\left(-i\frac{\hat{S}_z}{\hbar}\phi\right). \tag{8.15}$$

We now want to calculate how the expectation value of and operator, say $\hat{S}_x$ changes due to this rotation.

$$\langle\alpha|\hat{S}_x|\alpha\rangle \rightarrow \langle\alpha|\hat{S}_x|\alpha\rangle_R$$
$$= \langle\alpha|e^{i\frac{\hat{S}_z\phi}{\hbar}}\hat{S}_x e^{-i\frac{\hat{S}_z\phi}{\hbar}}|\alpha\rangle. \tag{8.16}$$

We can evaluate Eq. (8.16) in two ways.

(i). METHOD 1.
   We use the expression Eq. (8.12) for $\hat{S}_x$. Then

$$
\begin{aligned}
e^{i\frac{\hat{S}_z\phi}{\hbar}}\hat{S}_x e^{-i\frac{\hat{S}_z\phi}{\hbar}} &= \frac{\hbar}{2}e^{i\frac{\hat{S}_z\phi}{\hbar}}[|+\rangle\langle-|+|-\rangle\langle+|]e^{-i\frac{\hat{S}_z\phi}{\hbar}} \\
&= \frac{\hbar}{2}\left[e^{\frac{i\phi}{2}}|+\rangle\langle-|e^{\frac{i\phi}{2}}+e^{\frac{-i\phi}{2}}|-\rangle\langle+|e^{\frac{-i\phi}{2}}\right] \\
&= \frac{\hbar}{2}[(\cos\phi+i\sin\phi)|+\rangle\langle-|+(\cos\phi-i\sin\phi)|-\rangle\langle+|] \\
&= \frac{\hbar}{2}(\cos\phi[|+\rangle\langle-|+|-\rangle\langle+|]+i\sin\phi[|+\rangle\langle-|-|-\rangle\langle+|]) \\
&= \cos\phi\hat{S}_x-\sin\phi\hat{S}_y. \tag{8.17}
\end{aligned}
$$

(ii). METHOD 2.
   We can use Eq. (6.143), which follows from Baker-Campbell-Hausdorff lemma, and get

$$
\begin{aligned}
e^{i\frac{\hat{S}_z}{\hbar}\phi}\hat{S}_x e^{-i\frac{\hat{S}_z}{\hbar}\phi} &= \hat{S}_x+\left(\frac{i\phi}{\hbar}\right)[\hat{S}_z,\hat{S}_x]+\frac{1}{2!}\left(\frac{i\phi}{\hbar}\right)^2[\hat{S}_z,[\hat{S}_z,\hat{S}_x]] \\
&\quad +\frac{1}{3!}\left(\frac{i\phi}{\hbar}\right)^3[\hat{S}_z,[\hat{S}_z,[\hat{S}_z,\hat{S}_x]]]+\cdots \\
&= \cos\phi\hat{S}_x-\sin\phi\hat{S}_y. \tag{8.18}
\end{aligned}
$$

Thus

$$
\begin{align}
\langle \hat{S}_x \rangle &\rightarrow \langle \alpha | \hat{S}_x | \alpha \rangle_R = \cos\phi \langle \hat{S}_x \rangle - \sin\phi \langle \hat{S}_y \rangle, \tag{8.19} \\
\langle \hat{S}_y \rangle &\rightarrow \langle \alpha | \hat{S}_y | \alpha \rangle_R = \sin\phi \langle \hat{S}_x \rangle + \cos\phi \langle \hat{S}_y \rangle, \tag{8.20} \\
\langle \hat{S}_z \rangle &\rightarrow \langle \alpha | \hat{S}_z | \alpha \rangle_R = \langle \hat{S}_z \rangle. \tag{8.21}
\end{align}
$$

The expectation value of $\hat{S}_z$ remains unchanged as $\hat{S}_z$ commutes with $\mathcal{D}_z(\phi)$.

Eq. (8.19), Eq. (8.20) and Eq. (8.21) show that the rotation operator Eq. (8.15) when applied to the state ket rotates the expectation value of $\hat{S}$ around the z-axis by an angle ϕ. The expectation values of the spin operators behave as if they were classical vectors under rotation.

8.4   Pauli Two Component Spinor Formalism

The two component spinor formalism introduced by Pauli in 1926 make the manipulations with the state kets of spin-1/2 system very convenient. We already know how a ket (bra) can be represented by a column (row) matrix. One has only to arrange the expansion coefficients in terms of a certain specified set of base kets into a column (row) matrix. In the case of Spin $= \frac{1}{2}$, we have

$$
|+\rangle \Rightarrow \begin{pmatrix} 1 \\ 0 \end{pmatrix} \equiv \chi_+ \qquad |-\rangle \Rightarrow \begin{pmatrix} 0 \\ 1 \end{pmatrix} \equiv \chi_-, \tag{8.22}
$$

$$
\langle +| \Rightarrow \begin{pmatrix} 1 & 0 \end{pmatrix} \equiv \chi_+^\dagger \qquad \langle -| \Rightarrow \begin{pmatrix} 0 & 1 \end{pmatrix} \equiv \chi_-^\dagger. \tag{8.23}
$$

for the base kets and bras and an arbitrary ket $|\alpha\rangle$ in this space is given by

$$
|\alpha\rangle = |+\rangle\langle +|\alpha\rangle + |-\rangle\langle -|\alpha\rangle \Rightarrow \begin{pmatrix} \langle +|\alpha\rangle \\ \langle -|\alpha\rangle \end{pmatrix}, \quad \text{and} \tag{8.24}
$$

$$
\langle \alpha| = \langle \alpha|+\rangle\langle +| + \langle \alpha|-\rangle\langle -| \Rightarrow \begin{pmatrix} \langle \alpha|+\rangle & \langle \alpha|-\rangle \end{pmatrix}. \tag{8.25}
$$

The column matrix Eq. (8.24) is referred to as a two component *spinor* and is written as

$$
\chi = \begin{pmatrix} \langle +|\alpha\rangle \\ \langle -|\alpha\rangle \end{pmatrix} = \begin{pmatrix} C_+ \\ C_- \end{pmatrix} = C_+\chi_+ + C_-\chi_-, \tag{8.26}
$$

where C_+ and C_- are in general complex numbers. Similarly we have

$$
\chi^\dagger = \begin{pmatrix} \langle \alpha|+\rangle & \langle \alpha|-\rangle \end{pmatrix} = \begin{pmatrix} C_+^* & C_-^* \end{pmatrix}. \tag{8.27}
$$

The matrix elements $\langle \pm | \hat{S}_k | + \rangle$ and $\langle \pm | \hat{S}_k | - \rangle$ apart from a factor of $\hbar/2$ are to be equal to those of 2×2 matrices σ_k, known as *Pauli matrices*

$$\langle \pm | \hat{S}_k | + \rangle \equiv \frac{\hbar}{2} (\sigma_k)_{\pm,+}, \quad \langle \pm | \hat{S}_k | - \rangle \equiv \frac{\hbar}{2} (\sigma_k)_{\pm,-}. \tag{8.28}$$

Then the expectation value $\langle \hat{S}_k \rangle$ can be written as

$$\langle \hat{S}_k \rangle = \langle \alpha | \hat{S}_k | \alpha \rangle = \sum_{a_1=+,-} \sum_{a_2=+,-} \langle \alpha | a_1 \rangle \langle a_1 | \hat{S}_k | a_2 \rangle \langle a_2 | \alpha \rangle$$

$$= \frac{\hbar}{2} \chi^\dagger \sigma_k \chi. \tag{8.29}$$

From the matrices for $\hat{S}_z$, $\hat{S}_x$ and $\hat{S}_y$ in Eq. (8.3), Eq. (8.6), Eq. (8.7) and Eq. (8.28) we can write the *Pauli Matrices* as

$$\sigma_1 = \begin{pmatrix} 0 & 1 \\ 1 & 0 \end{pmatrix}, \quad \sigma_2 = \begin{pmatrix} 0 & -i \\ i & 0 \end{pmatrix}, \quad \sigma_3 = \begin{pmatrix} 1 & 0 \\ 0 & -1 \end{pmatrix}. \tag{8.30}$$

where the subscripts 1, 2, 3 refer to x, y, z respectively.

The Pauli matrices σ_1, σ_2, σ_3 satisfy the following

$$\sigma_i^2 = \hat{I}, \tag{8.31}$$

$$\sigma_i \sigma_j + \sigma_j \sigma_i = 0 \text{ for } i \neq j, \tag{8.32}$$

where the right hand side of Eq. (8.31) is a 2×2 identity matrix. These relations are equivalent to the anticommutation relations

$$\{\sigma_i, \sigma_j\} = 2\delta_{i,j}. \tag{8.33}$$

Also we have

$$[\sigma_i, \sigma_j] = 2i\epsilon_{ijk}\sigma_k, \tag{8.34}$$

which is the commutation relation. Combining Eq. (8.33) and Eq. (8.34) we get the set of relations

$$\sigma_1\sigma_2 = -\sigma_2\sigma_1 = i\sigma_3, \tag{8.35}$$

$$\sigma_2\sigma_3 = -\sigma_3\sigma_2 = i\sigma_1, \tag{8.36}$$

$$\sigma_3\sigma_1 = -\sigma_1\sigma_3 = i\sigma_2. \tag{8.37}$$

We further note that

$$\sigma_i^\dagger = \sigma_i, \tag{8.38}$$

$$\det \sigma_i = -1, \tag{8.39}$$

$$Tr(\sigma_i) = 0. \tag{8.40}$$

Now consider $\boldsymbol{\sigma} \cdot \mathbf{a}$ where $\mathbf{a}$ is a vector in 3 dimensions:

$$\boldsymbol{\sigma} \cdot \mathbf{a} = \sum_i \sigma_i a_i = \begin{pmatrix} a_3 & a_1 - ia_2 \\ a_1 + ia_2 & -a_3 \end{pmatrix}, \tag{8.41}$$

which is thus a 2×2 matrix. We now consider a very important identity

$$(\boldsymbol{\sigma} \cdot \mathbf{a})(\boldsymbol{\sigma} \cdot \mathbf{b}) = \mathbf{a} \cdot \mathbf{b} + i\boldsymbol{\sigma} \cdot (\mathbf{a} \times \mathbf{b}). \tag{8.42}$$

Problem 8.3 Use the commutation and the anticommutation relations Eq. (8.33) and Eq. (8.34) and prove the identity Eq. (8.42).

For real components of $\mathbf{a}$ we get

$$(\boldsymbol{\sigma} \cdot \mathbf{a})^2 = |\mathbf{a}|^2. \tag{8.43}$$

Problem 8.4 Also from Eq. (8.43) show that

$$(\boldsymbol{\sigma} \cdot \hat{\mathbf{n}})^n = \begin{cases} \hat{I} & \text{for} \quad n = \text{even} \\ \boldsymbol{\sigma} \cdot \hat{\mathbf{n}} & \text{for} \quad n = \text{odd} \end{cases} \tag{8.44}$$

Chapter 9

Addition of Angular Momenta

9.1 Addition of Two Angular Momenta $\hat{\mathbf{J}}_1$ and $\hat{\mathbf{J}}_2$

We shall treat the simplest addition problem, namely that of adding two commuting angular momenta $\hat{\mathbf{J}}_1$ and $\hat{\mathbf{J}}_2$. Thus

$$\hat{\mathbf{J}} = \hat{\mathbf{J}}_1 + \hat{\mathbf{J}}_2, \tag{9.1}$$

where $\mathbf{J}_1$ and $\hat{\mathbf{J}}_2$ are any two angular momenta corresponding respectively to the independent subsystems S_1 and S_2 or sets of dynamical variables 1 and 2.

$|j_1, m_1\rangle$ is the normalized simultaneous eigenvector of $\hat{J}_1^2$ and $\hat{J}_{1z}$ and we have

$$\hat{J}_1^2|j_1, m_1\rangle = j_1(j_1 + 1)\hbar^2|j_1, m_1\rangle, \tag{9.2}$$
$$\hat{J}_{1z}|j_1, m_1\rangle = m_1\hbar|j_1, m_1\rangle. \tag{9.3}$$

Similarly for $\hat{J}_2^2$ and $\hat{J}_{2z}$ we have

$$\hat{J}_2^2|j_2, m_2\rangle = j_2(j_2 + 1)\hbar^2|j_2, m_2\rangle, \tag{9.4}$$
$$\hat{J}_{2z}|j_2, m_2\rangle = m_2\hbar|j_2, m_2\rangle. \tag{9.5}$$

A normalized simultaneous eigenvector of $\hat{J}_1^2$, $\hat{J}_2^2$, $\hat{J}_{1z}$ and $\hat{J}_{2z}$ belonging respectively to the eigenvalues $j_1(j_1 + 1)\hbar^2$, $j_2(j_2 + 1)\hbar^2$, $m_1\hbar$ and $m_2\hbar$ is then given by the direct product

$$|j_1, j_2; m_1, m_2\rangle = |j_1, m_1\rangle|j_2, m_2\rangle. \tag{9.6}$$

For a fixed value of j_1, m_1 can take one of the $2j_1 + 1$ values $-j_1$, $-j_1 + 1$, $\cdots$, $j_1 - 1$, j_1 and for a fixed value of j_2 the $2j_2 + 1$ allowed values of m_2 are $-j_2$, $-j_2 + 1$, $\cdots$, $j_2 - 1$, j_2. Hence for given values of j_1 and j_2 there are $(2j_1 + 1)(2j_2 + 1)$

direct products Eq. (9.6) which form a complete orthonormal set, *i.e.* a basis in the product space of the combined system $(1 + 2)$.

We rewrite Eq. (9.6) in a more instructive way by denoting $\psi_{j_1,m_1}(1)$ instead of $|j_1, m_1\rangle$ and $\psi_{j_2,m_2}(2)$ instead of $|j_2, m_2\rangle$. Then the normalized simultaneous eigenfunctions of $\hat{J}_1^2$, $\hat{J}_2^2$, $\hat{J}_{1z}$ and $\hat{J}_{2z}$ corresponding to the eigenvalues $j_1(j_1+1)\hbar^2$, $j_2(j_2+1)\hbar^2$, $m_1\hbar$ and $m_2\hbar$ are

$$\psi_{j_1,j_2;m_1,m_2}(1,2) = \psi_{j_1,m_1}(1)\psi_{j_2,m_2}(2). \tag{9.7}$$

Now

$$\begin{aligned}
\hat{J}_z\psi_{j_1,j_2;m_1,m_2}(1,2) &= \left(\hat{J}_{1z} + \hat{J}_{2z}\right)\psi_{j_1,m_1}(1)\psi_{j_2,m_2}(2) \\
&= \left[\hat{J}_{1z}\psi_{j_1,m_1}(1)\right]\psi_{j_2,m_2}(2) + \psi_{j_1,m_1}(1)\left[\hat{J}_{2z}\psi_{j_2,m_2}(2)\right] \\
&= m_1\hbar\psi_{j_1,m_1}(1)\psi_{j_2,m_2}(2) + m_2\hbar\psi_{j_1,m_1}(1)\psi_{j_2,m_2}(2) \\
&= (m_1 + m_2)\hbar\psi_{j_1,j_2;m_1,m_2}(1,2), \tag{9.8}
\end{aligned}$$

which shows that $\psi_{j_1,j_2;m_1,m_2}(1,2)$ is also an eigenfunction of $\hat{J}_z$ corresponding to the eigenvalue $(m_1 + m_2)\hbar$. In abstract notation

$$\hat{J}_z|j_1, j_2; m_1, m_2\rangle = (m_1 + m_2)\hbar|j_1, j_2; m_1, m_2\rangle. \tag{9.9}$$

Now we consider the operator $\hat{J}^2$

$$\begin{aligned}
\hat{J}^2 &= \left(\hat{\mathbf{J}}_1 + \hat{\mathbf{J}}_2\right)^2 = \hat{\mathbf{J}}_1^2 + \hat{\mathbf{J}}_2^2 + \hat{\mathbf{J}}_1 \cdot \hat{\mathbf{J}}_2 + \hat{\mathbf{J}}_2 \cdot \hat{\mathbf{J}}_1 \\
&= \hat{\mathbf{J}}_1^2 + \hat{\mathbf{J}}_2^2 + 2\hat{\mathbf{J}}_1 \cdot \hat{\mathbf{J}}_2. \tag{9.10}
\end{aligned}$$

Since all the components of $\hat{\mathbf{J}}_1$ commute with all those of $\hat{\mathbf{J}}_2$ and

$$\hat{\mathbf{J}}_1 \cdot \hat{\mathbf{J}}_2 = \hat{J}_{1x}\hat{J}_{2x} + \hat{J}_{1y}\hat{J}_{2y} + \hat{J}_{1z}\hat{J}_{2z}. \tag{9.11}$$

Because $\hat{J}_{1z}$ does not commute with $\hat{J}_{1x}$ and $\hat{J}_{1y}$, $\hat{J}^2$ does not commute with $\hat{J}_{1z}$. Similarly $\hat{J}^2$ does not commute with $\hat{J}_{2z}$. Consequently the simultaneous eigenfunctions of $\hat{J}^2$ and $\hat{J}_z$ are eigenfunctions of $\hat{J}_1^2$ and $\hat{J}_2^2$ but not in general of $\hat{J}_{1z}$ and $\hat{J}_{2z}$. Thus there are two complete but distinct descriptions of the system

(i). In term of the eigenfunctions of $\hat{J}_1^2$, $\hat{J}_2^2$, $\hat{J}_{1z}$ and $\hat{J}_{2z}$, given by Eq. (9.7) or

(ii). In terms of of the eigenfunctions of $\hat{J}_1^2$, $\hat{J}_2^2$, $\hat{J}^2$ and $\hat{J}_z$. This latter we denote by the normalized wavefunctions $\phi_{j_1,j_2}^{j,m}(1,2)$, and we have

$$\hat{J}^2\phi_{j_1,j_2}^{j,m}(1,2) = j(j+1)\hbar^2\phi_{j_1,j_2}^{j,m}(1,2), \quad \text{and} \tag{9.12}$$

$$\hat{J}_z\phi_{j_1,j_2}^{j,m}(1,2) = m\hbar\phi_{j_1,j_2}^{j,m}(1,2). \tag{9.13}$$

Like the functions $\psi_{j_1,j_2;m_1.m_2}(1,2)$ the functions $\phi_{j_1,j_2}^{j,m}(1,2)$ form a complete orthonormal set and are another basis in the product space of the system $(1+2)$. These two basis sets are related by a unitary transformation which we discussed in § (2.2). Since the identity operator is defined by

$$\hat{I} = \sum_{m_1,m_2} |j_1,j_2;m_1,m_2\rangle\langle j_1,j_2;m_1,m_2|, \tag{9.14}$$

$$\therefore \ |j_1,j_2;j,m\rangle = \sum_{m_1,m_2} |j_1,j_2;m_1,m_2\rangle\langle j_1,j_2;m_1,m_2|j_1,j_2;j,m\rangle, \tag{9.15}$$

$$\text{or } \phi_{j_1,j_2}^{j,m}(1,2) = \sum_{m_1,m_2} \langle j_1,j_2;m_1,m_2|j_1,j_2;j,m\rangle\psi_{j_1,j_2;m_1,m_2}(1,2). \tag{9.16}$$

The elements of the transformation matrix $\langle j_1,j_2;m_1,m_2|j_1,j_2;j,m\rangle$ are the *Clebsch-Gordan coefficients* or *Vector Addition coefficients*.

There are many important properties of Clebsch- Gordan coefficients (known in short as C-G coefficients), that we are now ready to study.

First the coefficients vanish unless

$$m = m_1 + m_2. \tag{9.17}$$

To prove this, we first note that

$$\left(\hat{J}_z - \hat{J}_{1z} - \hat{J}_{2z}\right)|j_1,j_2;j,m\rangle = 0. \tag{9.18}$$

Multiplying this with $\langle j_1,j_2;m_1,m_2|$ on the left, we obtain

$$(m - m_1 - m_2)\langle j_1,j_2;m_1,m_2|j_1,j_2;j,m\rangle = 0, \tag{9.19}$$

which proves our assertion. (The Dirac notation is admirably powerful!)

Second, the coefficients vanish unless

$$|j_1 - j_2| \leq j \leq j_1 + j_2. \tag{9.20}$$

This property appears obvious from the vector model of angular momentum addition, where we may visualize $\mathbf{J}$ to be the vectorial sum of $\mathbf{J}_1$ and $\mathbf{J}_2$. It can also be checked by showing that if Eq. (9.20) holds then the dimensionality of the space spanned by $\{|j_1,j_2;m_1,m_2\rangle\}$ is the same as that of the space spanned by $\{|j_1,j_2;j,m\rangle\}$. For (m_1,m_2) way of counting we get

$$N = (2j_1 + 1)(2j_2 + 1), \tag{9.21}$$

because for given j_1 there are $(2j_1 + 1)$ possible values of m_1; similarly for a given value of j_2. As for (j,m) way of counting, we note that for each j, there are $(2j+1)$

m values and according to Eq. (9.20), j itself runs from $j_1 - j_2$ to $j_1 + j_2$, where we have assumed without loss of generality, that $j_1 \geq j_2$. We therefore obtain

$$
\begin{aligned}
N &= \sum_{j=j_1-j_2}^{j_1+j_2} (2j+1) \\
&= 2 \sum_{j=j_1-j_2}^{j_1+j_2} j + \sum_{j=j_1-j_2}^{j_1+j_2} 1 \\
&= 2 \cdot \frac{(2j_2+1)\left(\overline{j_1+j_2} + \overline{j_1-j_2}\right)}{2} + (2j_2+1) \\
&= (2j_1+1)(2j_2+1).
\end{aligned}
\tag{9.22}
$$

Thus Eq. (9.20) is consistent. We thus have

$$
j = |j_1 - j_2|, \ |j_1 - j_2| + 1, \ \cdots, \ j_1 + j_2 - 1, \ j_1 + j_2.
\tag{9.23}
$$

The C-G coefficients form a unitary matrix whose matrix elements are taken to be real by convention. An immediate consequence of this is that the inverse coefficient $\langle j_1, j_2; j, m | j_1, j_2; m_1, m_2 \rangle$ is the same as $\langle j_1, j_2; m_1, m_2 | j_1, j_2; j, m \rangle$ itself.

A real unitary matrix is orthogonal, so we have the orthogonality condition

$$
\sum_j \sum_m \langle j_1, j_2; m_1, m_2 | j_1, j_2; j, m \rangle \langle j_1, j_2; m_1', m_2' | j_1, j_2; j, m \rangle = \delta_{m_1, m_1'} \delta_{m_2, m_2'}.
\tag{9.24}
$$

Likewise

$$
\sum_{m_1} \sum_{m_2} \langle j_1, j_2; m_1, m_2 | j_1, j_2; j, m \rangle \langle j_1, j_2; m_1, m_2 | j_1, j_2; j', m' \rangle = \delta_{j,j'} \delta_{m,m'}.
\tag{9.25}
$$

As a special case of this we may set $j' = j$, $m' = m = m_1 + m_2$. We then obtain

$$
\sum_{m_1} \sum_{m_2} |\langle j_1, j_2; m_1, m_2 | j_1, j_2; j, m = m_1 + m_2 \rangle|^2 = 1,
\tag{9.26}
$$

which is just the normalization condition of $|j_1, j_2; j, m\rangle$.

G. Racah gave an expression for the C-G coefficients which are very convenient for actual calculations

$$
\langle j_1, j_2; m_1, m_2 | j_1, j_2; j, m \rangle = \delta_{m=m_1+m_2} \sqrt{\frac{(2j+1)(j_1+j_2-j)!(j+j_1-j_2)!}{(j+j_1+j_2+1)!}} \times
$$

$$
\sqrt{(j_1+m_1)!(j_1-m_1)!(j_2+m_2)!(j_2-m_2)!} \times
$$

$$
\sqrt{(j+m)!(j-m)!} \sum_s \frac{(-1)^s}{s!(j_1+j_2-j-s)!} \times
$$

$$\frac{1}{(j_1 - m_1 - s)! \, (j - j_2 + m_1 + s)!} \times$$
$$\frac{1}{(j_2 + m_2 - s)! \, (j - j_1 - m_2 + s)!}. \tag{9.27}$$

C-G coefficients can also be written in terms of Wigner $3 - j$ symbols. We give here the first two examples of the Clebsch-Gordan coefficients.

Table 9.1 Table of non-vanishing Clebsch-Gordan coefficients: $\langle j_1, \frac{1}{2}; m_1, m_2 | j_1, \frac{1}{2}; j, m = m_1 + m_2 \rangle$.

$j \downarrow$	$m_2 = +\frac{1}{2}$	$m_2 = -\frac{1}{2}$
$j_1 + \frac{1}{2}$	$+\sqrt{\frac{j_1 + m_1 + 1}{2j_1 + 1}}$	$+\sqrt{\frac{j_1 - m_1 + 1}{2j_1 + 1}}$
$j_1 - \frac{1}{2}$	$-\sqrt{\frac{j_1 - m_1}{2j_1 + 1}}$	$+\sqrt{\frac{j_1 + m_1}{2j_1 + 1}}$

Table 9.2 Table of non-vanishing Clebsch-Gordan coefficients: $\langle j_1, 1; m_1, m_2 | j_1, 1; j, m = m_1 + m_2 \rangle$.

$j \downarrow$	$m_2 = +1$	$m_2 = 0$	$m_2 = -1$
$j_1 + 1$	$+\sqrt{\frac{(j_1 + m_1 + 1)(j_1 + m_1 + 2)}{(2j_1 + 1)(2j_1 + 2)}}$	$+\sqrt{\frac{(j_1 - m_1 + 1)(j_1 + m_1 + 1)}{(2j_1 + 1)(j_1 + 1)}}$	$+\sqrt{\frac{(j_1 - m_1 + 1)(j_1 - m_1 + 2)}{(2j_1 + 1)(2j_1 + 2)}}$
j_1	$-\sqrt{\frac{(j_1 + m_1 + 1)(j_1 - m_1)}{2j_1 (j_1 + 1)}}$	$+\frac{m_1}{\sqrt{j_1 (j_1 + 1)}}$	$+\sqrt{\frac{(j_1 - m_1 + 1)(j_1 + m_1)}{2j_1 (j_1 + 1)}}$
$j_1 - 1$	$+\sqrt{\frac{(j_1 - m_1 - 1)(j_1 - m_1)}{2j_1 (2j_1 + 1)}}$	$-\sqrt{\frac{(j_1 - m_1)(j_1 + m_1)}{j_1 (2j_1 + 1)}}$	$+\sqrt{\frac{(j_1 + m_1)(j_1 + m_1 - 1)}{2j_1 (2j_1 + 1)}}$

9.2 Addition of Orbital Angular Momentum and Spin of a Particle

As a first example, we consider a particle of spin s. Let $\hat{\mathbf{L}}$ be the orbital angular mpmentum operator and $\hat{\mathbf{S}}$ its spin operator. The total angular operator of the particle is therefore $\hat{\mathbf{J}} = \hat{\mathbf{L}} + \hat{\mathbf{S}}$. We denote by m_l, m_s and m the quantum numbers corresponding to the operators $\hat{L}_z$, $\hat{S}_z$ and $\hat{J}_z$ respectively. In the position representation the simultaneous eigenfunctions of the operators $\hat{L}^2$ and $\hat{L}_z$ are the Spherical Harmonics $Y_{l,m_l}(\theta, \phi)$, with $l = 0,\ 1,\ 2,\ \cdots$ and $m_l = -l,\ -l+1,\ \cdots,\ l-1,\ l$. The simultaneous eigenfunctions of the operators $\hat{S}^2$ and $\hat{S}_z$ are the spin functions χ_{s,m_s} (with $m_s = -s,\ -s+1,\ \cdots,\ s-1,\ s$) which is represented by column matrices with $(2s + 1)$ rows with zeroes in all rows except one. Hence the simultaneous eigenfunctions of the operators $\hat{L}^2$, $\hat{S}^2$, $\hat{L}_z$ and $\hat{S}_z$ are represented by the product

$$\psi_{l,s;m_l,m_s} \; = \; Y_{l,m_l}(\theta, \phi)\,\chi_{s,m_s}. \tag{9.28}$$

The allowed values of the total angular momentum quantum number j of the particle are

$$j \; = \; |l - s|, \; |l - s| + 1, \; \cdots, \; l + s - 1, \; l + s. \tag{9.29}$$

Denoting the simultaneous normalized eigenfunctions of the operators $\hat{L}^2$, $\hat{S}^2$, $\hat{J}^2$ and $\hat{J}_z$ by $\mathcal{Y}_{l,s}^{j,m}$ we see that

$$
\begin{aligned}
\mathcal{Y}_{l,s}^{j,m} \; &= \; \sum_{m_l,m_s} \langle l, s; m_l.m_s | l, s; j, m \rangle \psi_{l,s;m_l,m_s} \\
&= \; \sum_{m_l,m_s} \langle l, s; m_l.m_s | l, s; j, m \rangle Y_{l,m_l}(\theta, \phi)\,\chi_{s,m_s}
\end{aligned}
\tag{9.30}
$$

For a particle of spin $s = \frac{1}{2}$ we see from Eq. (9.29) that for a given value of the orbital angular momentum quantum number l the total angular momentum quantum number j can take values

$$j \; = \; l - \frac{1}{2}, \; l + \frac{1}{2}, \tag{9.31}$$

except when $l = 0$ (S state) in which case the only allowed value is $j = \frac{1}{2}$. By using the C-G coefficients in Table (9.1) we can write

$$
\mathcal{Y}_{l,\frac{1}{2}}^{l\pm\frac{1}{2},m} \; = \;
\begin{pmatrix}
\pm\sqrt{\dfrac{l\pm m+\frac{1}{2}}{2l+1}}\,Y_{l,m-\frac{1}{2}}(\theta, \phi) \\[2ex]
\sqrt{\dfrac{l\mp m+\frac{1}{2}}{2l+1}}\,Y_{l,m+\frac{1}{2}}(\theta, \phi)
\end{pmatrix}.
\tag{9.32}
$$

9.3   Addition of Two Spins

The second example we consider is that of two particles whose spin operators are $\hat{\mathbf{S}}_1$ and $\hat{\mathbf{S}}_2$ respectively. Then

$$\hat{\mathbf{S}} \; = \; \hat{\mathbf{S}}_1 + \hat{\mathbf{S}}_2, \tag{9.33}$$

where $\hat{\mathbf{S}}$ is the total spin angular momentum. If the two particles have spin $\frac{1}{2}$ each, then the combined spin space has 4 dimensions. The simultaneous eigenfunctions of the operators $\hat{S}_1^2$ and $\hat{S}_{1z}$ for the particle 1 are two basic spinors $\alpha(1)$ and $\beta(1)$

corresponding respectively to 'spin up' ($m_{s_1} = +\frac{1}{2}$) and 'spin down' ($m_{s_1} = -\frac{1}{2}$) for that particle. Similarly the eigenfunctions of $\hat{S}_2^2$ and $\hat{S}_{2z}$ for particle 2 are the two basic spinors $\alpha\,(2)$ and $\beta\,(2)$ corresponding respectively to 'spin up' ($m_{s_2} = +\frac{1}{2}$) and 'spin down' ($m_{s_2} = -\frac{1}{2}$) for the second particle. The direct product eigenfunctions $\psi_{j_1,j_2;m_1,m_2}\,(1,2)$ are therefore in the present case the four spin functions

$$\alpha\,(1)\,\alpha\,(2)\,,\ \ \alpha\,(1)\,\beta\,(2)\,,\ \ \beta\,(1)\,\alpha\,(1)\,,\ \ \beta\,(1)\,\beta\,(2)\,, \tag{9.34}$$

which constitute a basis in the four dimensional system. If we denote by $M_s\hbar$ the eigenvalues of the operator $\hat{S}_z$, we see that $M_s = m_{s_1} + m_{s_2}$, so that the four eigenfunctions Eq. (9.34) correspond respectively to the values $M_s = 1,\ 0,\ 0,\ -1$. The allowed values of the total spin quantum number S are given from Eq. (9.23)

$$S\ =\ 0,\ 1. \tag{9.35}$$

If we denote by χ_{S,M_S} the simultaneous normalized eigenfunctions of $\hat{S}_1^2,\ \hat{S}_2^2,\ \hat{S}^2$ and $\hat{S}_z$ are given by Eq. (9.15) and Eq. (9.34), for $S = 0$

$$\chi_{0,0}\ =\ \frac{1}{\sqrt{2}}\left[\alpha\,(1)\,\beta\,(2) - \beta\,(1)\,\alpha\,(2)\right], \tag{9.36}$$

which is called *singlet spin state*. This has been constructed so that it is *antisymmetric* in the interchange of the spin coordinates of the two particles. For $S = 1$, we have

$$\chi_{1,1}\ =\ \alpha\,(1)\,\alpha\,(2)\,, \tag{9.37}$$

$$\chi_{1,0}\ =\ \frac{1}{\sqrt{2}}\left[\alpha\,(1)\,\beta\,(2) + \beta\,(1)\,\alpha\,(2)\right], \tag{9.38}$$

$$\chi_{1,-1}\ =\ \beta\,(1)\,\beta\,(2)\,. \tag{9.39}$$

These are said to form a *spin triplet state*. These three states, which are *symmetric* in the interchange of the spin coordinates of the two particles, are eigenstates of $\hat{S}_z$ corresponding respectively to the values of $M_S = +1,\ 0,\ -1$.

For an example, the lowest state of the helium atom which contains two electrons is a singlet state ($S = 0$) while the excited states can be either singlet or triplet states.

Chapter 10

Applications II

In this chapter we would consider motion in three dimensions. In § (10.1) we would study *Hydrogen Atom.* And in the next section we would consider motion of a charged particle in magnetic field, including *Landau Levels* and *Aharonov Bohm Effect.*

10.1 Hydrogen Atom

As an example of three dimensional problem we consider the hydrogen atom containing an electron of charge $-e$ interacting with a point nucleus of charge $+Ze$ by means of Coulomb potential

$$V(r) = -\frac{Ze^2}{r}, \tag{10.1}$$

where r is the distance between the two particles. We denote the electronic mass by m and the mass of the nucleus by M. It is convenient to use the centre of mass coordinate system as the potential is a function of the relative coordinate. The relative motion of the two particles is described by the Hamiltonian

$$H = \frac{\mathbf{p}^2}{2\mu} - \frac{Ze^2}{r} \tag{10.2}$$

in the centre of mass system (where the total momentum $\mathbf{P}$ of the atom equals zero), where $\mathbf{p}$ is the relative momentum and $\mu = \frac{mM}{m+M}$ is the reduced mass. The Schrödinger equation is given by

$$\left[-\frac{\hbar^2}{2\mu} \boldsymbol{\nabla}^2 - \frac{Ze^2}{r} \right] \psi(\mathbf{r}) = E\psi(\mathbf{r}). \tag{10.3}$$

We separate the solution of Eq. (10.3) in radial and angular parts

$$
\psi(r,\theta,\phi) = R_{E,l}(r)Y_{l,m}(\theta,\phi) \tag{10.4}
$$

$$
= \frac{u_{E,l}(r)}{r}Y_{l,m}(\theta,\phi)
$$

where $u_{E,l}(r) \to 0$ as $r \to 0$. We have

$$
\frac{d^2u_{E,l}(r)}{dr^2} + \frac{2\mu}{\hbar^2}\left[E + \frac{Ze^2}{r} - \frac{l(l+1)\hbar^2}{2\mu r^2}\right]u_{E,l}(r) = 0, \tag{10.5}
$$

$$
\text{or} \qquad \frac{d^2u_{E,l}(r)}{dr^2} + \frac{2\mu}{\hbar^2}\left[E - V_{\text{eff}}\right]u_{E,l}(r) = 0, \tag{10.6}
$$

where

$$
V_{\text{eff}}(r) = -\frac{Ze^2}{r} + \frac{l(l+1)\hbar^2}{2\mu r^2}. \tag{10.7}
$$

Since $V_{\text{eff}} \to 0$ at $r \to \infty$, the solution $u_{E,l}(r)$ for positive energy is oscillatory at infinity and will be acceptable eigenfunction. We shall have a continuous spectrum for $E > 0$ for unbound scattering states. These states are important in the analysis of collision phenomenon between electrons and ions.

Now the solution to the Eq. (10.3) has to satisfy certain boundary conditions relevant to the physical situation.

If for exampe $E < \lim_{r\to\infty} V(r)$ then the appropriate boundary condition in this case is

$$
\psi(r) \to 0, \quad \text{as } r \to \infty. \tag{10.8}
$$

This means that the particle is bounded or is localized within a finite region of space. From the theory of partial differential equations we know that the Eq. (10.3) subject to the boundary condition Eq. (10.8) will allow non-trivial solutions only for a discrete set of values of E. Thus the energy levels are quantized because of the boundary condition $u_{E,l}(0) = 0$, because $R_{E,l}(r)$ has to be finite at every point including at the origin. We introduce the dimensionless variable

$$
\rho = \sqrt{\frac{8\mu|E|}{\hbar^2}}r, \tag{10.9}
$$

and dimensionless energy

$$
\lambda = \left(\frac{Ze^2}{\hbar}\right)\sqrt{\frac{\mu}{2|E|}}
$$

$$
= Z\alpha\sqrt{\frac{\mu c^2}{2|E|}}, \tag{10.10}
$$

$$
\text{where} \quad \alpha = \frac{e^2}{\hbar c} \approx \frac{1}{137} \tag{10.11}
$$

is the *fine structure constant*. With these transformations we get

$$\left[\frac{d^2}{d\rho^2} - \frac{l(l+1)}{\rho^2} + \frac{\lambda}{\rho} - \frac{1}{4} \right] u_{E,l}(\rho) = 0. \tag{10.12}$$

To obtain the solution we first consider the asymptotic region $\rho \to 0$. We can expand the solution $u_{E,l}(\rho)$ in the vicinity of the origin as

$$u_{E,l}(\rho) = \rho^s \sum_{k=0}^{\infty} c_k \rho^k, \quad c_0 \neq 0. \tag{10.13}$$

Substituting Eq. (10.13) in Eq. (10.12) and equating the coefficient of the lowest power of ρ to zero, we get the *indicial equation*

$$s(s-1) - l(l+1) = 0, \tag{10.14}$$
$$\text{so} \quad s = l+1, \quad \text{or} \quad -l.$$

The solution $s = -l$ does not give the right behaviour as $\rho \to 0$. So we take $s = l+1$ and hence

$$u_{E,l} \sim \rho^{l+1}, \quad \text{as} \quad \rho \to 0. \tag{10.15}$$

Now for large ρ, the asymptotic equation is

$$\left[\frac{d^2}{d\rho^2} - \frac{1}{4} \right] u_{E,l}(\rho) = 0, \tag{10.16}$$

as $\rho \to \infty$, whose solutions are proportional to $\exp\left(\pm\frac{\rho}{2}\right)$ of which we retain $\exp\left(-\frac{\rho}{2}\right)$. Thus the solution is of the form

$$u_{E,l}(\rho) = \exp\left(-\frac{\rho}{2}\right) f(\rho). \tag{10.17}$$

Substituting Eq. (10.17) in Eq. (10.12) we see that $f(\rho)$ satisfies

$$\left[\frac{d^2}{d\rho^2} - \frac{d}{d\rho} - \frac{l(l+1)}{\rho^2} + \frac{\lambda}{\rho} \right] f(\rho) = 0. \tag{10.18}$$

We replace $f(\rho)$ by $g(\rho)$ such that

$$f(\rho) = \rho^{l+1} g(\rho) \tag{10.19}$$

so that the correct boundary condition Eq. (10.15) for $f(\rho)$ is satisfied as $\rho \to 0$.
Then $g(\rho)$ satisfies the following equation

$$\left[\rho \frac{d^2}{d\rho^2} + (2l+2-\rho)\frac{d}{d\rho} + (\lambda - l - 1) \right] g(\rho) = 0. \tag{10.20}$$

Now we expand $g(\rho)$ in an infinite series:

$$g(\rho) = \sum_{k=0}^{\infty} C_k \rho^k, \quad C_0 \neq 0. \tag{10.21}$$

We then have

$$\sum_{k=0}^{\infty} \left[k(k-1)C_k\rho^{k-1} + (2l+2-\rho)kC_k\rho^{k-1} + (\lambda-l-1)C_k\rho^k \right] = 0, \quad \text{or}$$

$$\sum_{k=0}^{\infty} \left\{ [k(k+1) + (2l+2)(k+1)]C_{k+1} + (\lambda-l-1-k)C_k \right\}\rho^k = 0. \tag{10.22}$$

So the coefficients C_k must satisfy the recursion relation

$$C_{k+1} = \frac{(k+l+1-\lambda)}{(k+1)(k+2l+2)}C_k. \tag{10.23}$$

If the series Eq. (10.21) does not terminate then for large k

$$\frac{C_{k+1}}{C_k} \sim \frac{1}{k}, \tag{10.24}$$

a ratio which is the same as that of the series $\rho^p \exp(\rho)$ where p has a finite value. In that case using Eq. (10.17) and Eq. (10.19) we deduce that $u_{E,l}(\rho)$ has large ρ an asymptotic behaviour

$$\lim_{\rho\to\infty} u_{E,l}(\rho) \sim \rho^{l+1+p}e^{\frac{\rho}{2}}, \tag{10.25}$$

which is not acceptable, because it blows up as $\rho \to \infty$.

The series Eq. (10.21) must therefore terminate, and $g(\rho)$ must be a polynomial in ρ. If the highest power of of ρ in $g(\rho)$ be ρ^{n_r}, where the radial quantum number $n_r = 1, 2, \cdots$ is a positive integer or zero, then the coefficient $C_{n_r+1} = 0$, and from the recursion relation Eq. (10.23)

$$\lambda = n_r + l + 1. \tag{10.26}$$

Introducing the *principal quantum number*

$$n = n_r + l + 1 \tag{10.27}$$

which is a positive integer $(n = 1, 2, \cdots)$ since both n_r and l can assume positive integral values or zero. Thus we see that the eigenvalue equation Eq. (10.12) corresponding to the bound state energy spectrum $(E < 0)$ are given by

$$\lambda = n. \tag{10.28}$$

Then fronm Eq. (10.10) we obtain the bound state energy eigenvalues

$$
\begin{aligned}
E_n &= -\frac{\mu}{2\hbar^2}\frac{\left(Ze^2\right)}{n^2} = -\frac{1}{2}\frac{\left(Ze\right)^2}{a_\mu}\frac{1}{n^2} \\
&= -\frac{1}{2}\mu c^2\frac{\left(Z\alpha\right)^2}{n^2}, \qquad \text{with} \quad n = 1,\, 2,\, \cdots \quad \text{and}
\end{aligned}
\tag{10.29}
$$

$$
\alpha = \frac{e^2}{\hbar c} \approx \frac{1}{137} \quad \text{is the fine structure constant and}
$$

$$
a_\mu = \frac{\hbar^2}{\mu e^2} = \frac{m}{\mu}a_0,
\tag{10.30}
$$

where a_μ is the modified Bohr radius for an atom of atomic number Z, and a_0 is the *Bohr radius*.

Thus the energy spectrum Eq. (10.29) agrees with the main feature of the experimental spectrum. This agreement is not perfect and various corrections like fine structure due to relativistic effect and electron spin, Lamb shift and hyperfine structure due to nuclear effects must be taken into acount in order to explain the details of the experimental energy spectrum.

DEGENERACY.

Since the energy eigenvalues E_n depend only on the principal quantum number n (Eq. (10.29)) they are degenerate with respect to the quantum numbers l and m. Indeed, for each value of n the orbital quantum number l may take on the values $0,\, 1,\, \cdots,\, n-1$ and for each values of l the magnetic quantum number m may take the $(2l+1)$ possible values $-l,\, -l+1,\, \cdots,\, +l-1,\, +l$. The total degeneracy of the bound state energy level E_n is given by

$$
\sum_{l=0}^{n-1}(2l+1) = 2\frac{n\left(n-1\right)}{2} + n = n^2.
\tag{10.31}
$$

The $(2l+1)$-fold degeneracy with respect to quantum number m is a feature for any central potential, occurring because of rotational symmetry. On the other hand, the degeneracy with respect to l is characteristic of the Coulomb potential and is called *accidental degeneracy*.

WAVE FUNCTIONS OF THE DISCRETE SPECTRUM

The hydrogenic wave functions may be written as

$$
\psi_{n,l,m}\left(r,\theta,\phi\right) = R_{n,l}\left(r\right)Y_{l,m}\left(\theta,\phi\right), \qquad \text{where}
\tag{10.32}
$$

$$
R_{n,l}\left(r\right) = Ne^{-\frac{\rho}{2}}\rho^l L_{n+l}^{2l+1}\left(\rho\right).
\tag{10.33}
$$

Here $L_{n+l}^{2l+1}(\rho)$ are the *associated Laguerre polynomials*. The normalization constant may be found by using the generating function to evaluate the integral

$$\int_0^\infty e^{-\rho}\rho^{2l}\left[L_{n+l}^{2l+1}(\rho)\right]^2\rho^2\,d\rho \;=\; \frac{2n\left[(n+l)!\right]^3}{(n-l-1)!}.\tag{10.34}$$

So the normalized radial functions for the bound states of Hydrogen atom are

$$R_{n,l}(r) \;=\; -\sqrt{\left(\frac{2Z}{na_\mu}\right)^3\frac{(n-l-1)!}{2n\left[(n+l)!\right]^3}}\;e^{-\frac{\rho}{2}}\rho^l L_{n+l}^{2l+1}(\rho),\quad \text{with}\tag{10.35}$$

$$\rho \;=\; \frac{2Z}{na_\mu}r,\quad \text{and}\quad a_\mu \;=\; \frac{\hbar^2}{\mu e^2}.\tag{10.36}$$

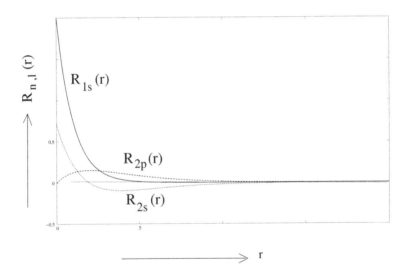

Figure 10.1: Radial wavefunctions $R_{1s}(r)$, $R_{2s}(r)$, $R_{2p}(r)$ of Hydrogen.

The first three radial functions Eq. (10.35) are given by

$$R_{1,0}(r) \;=\; R_{1s}(r) \;=\; 2\left(\frac{Z}{a_\mu}\right)^{\frac{3}{2}}\exp\left(-\frac{Zr}{a_\mu}\right)$$

$$R_{1,1}(r) \;=\; R_{1p}(r) \;=\; 2\left(\frac{Z}{2a_\mu}\right)^{\frac{3}{2}}\left(1-\frac{Zr}{2a_\mu}\right)\exp\left(-\frac{Zr}{2a_\mu}\right)$$

$$R_{1,1}(r) \;=\; R_{1p}(r) \;=\; \frac{1}{\sqrt{3}}\left(\frac{Z}{2a_\mu}\right)^{\frac{3}{2}}\left(\frac{Zr}{a_\mu}\right)\exp\left(-\frac{Zr}{a_\mu}\right)\tag{10.37}$$

and are shown in Fig. (10.1). Charge distribution within the atom is better depicted by the functions $|u_{n,l}(r)|^2$ from Eq. (10.4) connecting $u_{n,l}(r)$ and $R_{n,l}(r)$. These charge distributions correspondining to the 3 radial functions of Eq. (10.37) are shown in Fig. (10.2).

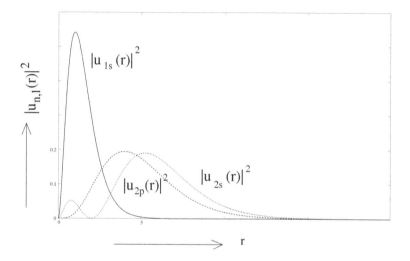

Figure 10.2: Charge Distributions $|u_{1s}(r)|^2$, $|u_{2s}(r)|^2$, $|u_{2p}(r)|^2$ of Hydrogen.

10.2 Charged Particle in Magnetic Field

(i) The Landau Levels

We consider the motion of an electron, for example in a uniform external magnetic field **B**. The Hamiltonian is

$$\hat{H} = \frac{1}{2m}\left(\hat{\mathbf{p}} + \frac{e\mathbf{A}}{c}\right)^2, \tag{10.38}$$

where the electronic charge is $-e$. We work in Coulomb gauge also called Landau gauge and take $\mathbf{B} = B\hat{\mathbf{k}}$, along te z-axis. We take

$$\mathbf{A} = \left(-By\hat{\mathbf{i}}, 0, 0\right), \tag{10.39}$$

so that $\nabla \cdot \mathbf{A} = 0$. (Coulomb or Landau gauge.)

Here $\mathbf{A}$ is the magnetic vector potential, $\mathbf{p}$ is the canonical momentum and $\boldsymbol{\pi} = \mathbf{p} + \frac{e\mathbf{A}}{c}$ is the kinetic momentum. Then

$$\hat{H} = \frac{1}{2m}\left[\left(\hat{p}_x - \frac{eBy}{c}\right)^2 + \hat{p}_y^2 + \hat{p}_z^2\right]. \tag{10.40}$$

Since $\hat{p}_x$ and $\hat{p}_z$ commute with $\hat{H}$, they are constants of motion:

$$\frac{d\hat{p}_x}{dt} = 0 = \frac{d\hat{p}_z}{dt}$$

and p_x and p_z are c-numbers we can then write the wavefunction as

$$\psi(x, y, z) = e^{ip_x x/\hbar} e^{ip_z z/\hbar} \phi(y). \tag{10.41}$$

Then $\phi(y)$ satisfies

$$\left[-\frac{\hbar^2}{2m}\frac{d^2}{dy^2} + \frac{1}{2}m\omega_c^2(y - y_0)^2 - E'\right]\phi(y) = 0 \tag{10.42}$$

$$\omega_c = \frac{eB}{mc}, \quad \text{is called the cyclotron frequency,} \tag{10.43}$$

$$E' = E - \frac{k_z^2\hbar^2}{2m}, \quad k_z\hbar = p_z \quad \text{and} \quad y_0 = \frac{p_x}{m\omega}.$$

Finally since for this one dimensional harmonic oscillator $E' = \left(n + \frac{1}{2}\right)\hbar\omega_c$

$$E_{n,k_z} = \left(n + \frac{1}{2}\right)\hbar\omega_c + \frac{k_z^2\hbar^2}{2m}. \tag{10.44}$$

is the energy of the electron in magnetic field. These are called the Landau levels, which are infinitely degenerate, since energy does not depend on p_x. In Fig. (10.3) we have shown the formation of Landau levels.

The wave function $\psi(x, y, z)$ is given in terms of the Hermite polynomials by

$$\psi_{n,k_z}(x, y, z) \propto e^{\frac{i}{\hbar}(p_x x + p_z z)} H_n(\alpha(y - y_0)) \exp\left[-\frac{1}{2}(\alpha(y - y_0))^2\right], \tag{10.45}$$

$$\text{where} \quad \alpha = \sqrt{\frac{m\omega_c}{\hbar}}$$

has the dimension of the inverse of a length.

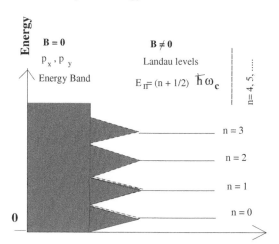

Figure 10.3: Collapse of continuum bands into Landau levels,

(ii)   The Aharanov Bohm Effect

In classical physics, only the electric and magnetic fields are of physical significance. The vector and scalar potentials are convenient mathematical devices to calculate the fields. The potentials can be redefined through a gauge transformation without changing the fields and hence without changing any physical laws. The use of vector potential in quantum mechanics has many far-reching consequences as was shown by Aharonov and Bohm (1959, 1961).

We consider (see Fig. (10.4) a particle of charge e going above or below a very long solenoid perpendicular to the plane of the paper carrying a current $\mathbf{j}$ and is surrounded by an impenetrable cylinder. Inside the cylinder is a magnetic field parallel to the axis of the cylinder, so the particle paths P_1 and P_2 enclose a magnetic flux.

We use Feynman path integral method of calculating probability amplitude. From classical mechanics the Lagrangian in the presence of magnetic field can be obtained from that in the absence of the magnetic field denoted by $L_{\text{cl}}^{(0)}$ as follows

$$L_{\text{cl}}^{(0)} = \frac{m}{2}\left(\frac{d\mathbf{x}}{dt}\right)^2 \longrightarrow L_{\text{cl}}^{(0)} + \frac{e}{c}\frac{d\mathbf{x}}{dt}\cdot\mathbf{A}. \qquad (10.46)$$

The corresponding change in action for some definite segment of path from (x_{n-1}, t_{n-1})

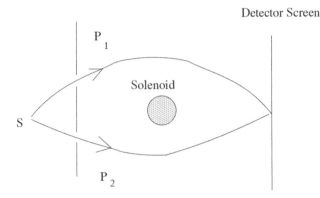

Figure 10.4: Arrangement for observing Aharonov-Bohm effect

to (x_n, t_n) is given by

$$S^{(0)}(n, n-1) \quad \longrightarrow \quad S^{(0)}(n, n-1) + \frac{e}{c} \int_{t_{n-1}}^{t_n} dt \frac{d\mathbf{x}}{dt} \cdot \mathbf{A}. \tag{10.47}$$

Now

$$\int_{t_{n-1}}^{t_n} dt \, \frac{d\mathbf{x}}{dt} \cdot \mathbf{A} \quad = \quad \int_{\mathbf{x}_{n-1}}^{\mathbf{x}_n} \mathbf{A} \cdot d\mathbf{s}, \tag{10.48}$$

where $d\mathbf{s}$ is the differential line element along the path segment. For the entire contribution from x_1 to x_N we have the following change

$$\prod_{2}^{N} \exp\left[i\frac{S^{(0)}(n, n-1)}{\hbar}\right] \quad \longrightarrow \quad \left(\prod_{2}^{N} \exp\left[i\frac{S^{(0)}(n, n-1)}{\hbar}\right]\right) \times$$
$$\exp\left(i\frac{e}{\hbar c} \int_{x_1}^{x_N} \mathbf{A} \cdot d\mathbf{s}\right). \tag{10.49}$$

Thus the path integral along $P_i, (i = 1, 2)$ acquires an extra factor

$$\exp\left(i\frac{e}{\hbar c} \int_{P_i} \mathbf{A} \cdot d\mathbf{s}\right). \tag{10.50}$$

For the entire transition amplitude $\langle x_N, t_N | x_1, t_1 \rangle$ (Eq. (5.25))

$$\int_{P_1} \mathcal{D}\left[\mathbf{x}(t)\right] \exp\left[i\frac{S^{(0)}(N, 1)}{\hbar}\right] + \int_{P_2} \mathcal{D}\left[\mathbf{x}(t)\right] \exp\left[i\frac{S^{(0)}(N, 1)}{\hbar}\right] \quad \longrightarrow$$

$$\int_{P_1} \mathcal{D}\left[\mathbf{x}\left(t\right)\right] \exp\left[i\frac{S^{(0)}\left(N,1\right)}{\hbar}\right] \times \exp\left[\left(i\frac{e}{\hbar c}\right)\int_{\mathbf{x_1}}^{\mathbf{x}_N} \mathbf{A}\cdot d\mathbf{s}\right]_{P_1} +$$

$$\int_{P_2} \mathcal{D}\left[\mathbf{x}\left(t\right)\right] \exp\left[i\frac{S^{(0)}\left(N,1\right)}{\hbar}\right] \times \exp\left[\left(i\frac{e}{\hbar c}\right)\int_{\mathbf{x_1}}^{\mathbf{x}_N} \mathbf{A}\cdot d\mathbf{s}\right]_{P_2}. \tag{10.51}$$

The probability being modulus squired of the entire transition amplitude and hence depends on the phase difference between the contributions from the paths P_1 and P_2. This phase difference due to the presence of $\mathbf{B}$ is then

$$\left[\left(\frac{e}{\hbar c}\right)\int_{\mathbf{x_1}}^{\mathbf{x}_N} \mathbf{A}\cdot d\mathbf{s}\right]_{P_1} - \left[\left(\frac{e}{\hbar c}\right)\int_{\mathbf{x_1}}^{\mathbf{x}_N} \mathbf{A}\cdot d\mathbf{s}\right]_{P_2} = \frac{e}{\hbar c}\oint \mathbf{A}\cdot d\mathbf{s}$$

$$= \frac{e}{\hbar c}\Phi_B, \tag{10.52}$$

where Φ_B is the magnetic flux crossing the surface bounded by P_1 and P_2 and there is a sinusoidal component in the probability for observing the particle on the detector screen with a period equal to

$$\frac{2\pi\hbar c}{|e|} = 4.135 \times 10^{-7} \text{ Gauss cm}^2. \tag{10.53}$$

This effect is purely quantum mechanical.

This was first observed by R. G. Chambers in 1960 using magnetic one-domain iron whiskers and also later by Tonomura *et al.* with the help of a superconducting film using electron holography.

Chapter 11

Symmetry in Quantum Mechanics

Symmetry is a quality attributed to the physical world that helps to simplify the study of physical systems as such. Thus we have seen in the study of hydrogen atom how symmetry consideration could reduce the number of meaningful variables needed to describe it from seven (6 coordinates and time t) to one, simply from symmetry considration.

11.1 Symmetry Principle and Conservation Laws

In this section we discuss the connection between symmetry principles and conservation laws. We designate the infinitesimal symmetry operator by $\hat{S}$ which can be written as

$$\hat{S} = \hat{I} - i\frac{\epsilon}{\hbar}\hat{G}, \tag{11.1}$$

where the Hermitian operator $\hat{G}$ is the generator of the symmetry operation and ϵ is the infinitesimal transformation being studied. If the Hamiltonian is invariant under $\hat{S}$ then

$$\hat{S}\hat{H} = \hat{H}\hat{S}, \tag{11.2}$$

$$\text{and} \quad \hat{G}\hat{H} = \hat{H}\hat{G} \tag{11.3}$$

$$\text{or} \quad \frac{d\hat{G}}{dt} = 0, \tag{11.4}$$

from Heisenberg equation of motion. Thus $\hat{G}$ is a *constant of motion*. This means that if $\hat{H}$ is invariant under translation (homogeneity of space) then linear momentum is a constant of motion. Similarly, the angular momentum is conserved if the Hamiltonian is rotationally invariant (isotropy of space). These are geometrical symmetries and finite operation can be achieved by successive infinitesimal transformations.

(i)   Symmetry and Degeneracy

If some symmetry operator $\hat{S}$ commute with the Hamiltonian $\hat{H}$ they have common eigenket.

$$\left[\hat{H}, \hat{S}\right] = 0. \tag{11.5}$$

If $|n\rangle$ is an eigenket of $\hat{H}$ with eigenvalue E_n, then

$$\hat{H}\left[\hat{S}|n\rangle\right] = \hat{S}\left[\hat{H}|n\rangle\right] = E_n\left[\hat{S}|n\rangle\right]. \tag{11.6}$$

Thus $\hat{S}|n\rangle$ is also an eigenket of $\hat{H}$ with the same eigenvalue. Thus $|n\rangle$ and $\hat{S}|n\rangle$ are degenerate. If $\hat{S}$ represents a continuous transformation characterized by a continuous parameter λ, then $\hat{S}(\lambda)|n\rangle$ and $|n\rangle$ represent different states belonging to the same energy E_n. All such states $\hat{S}(\lambda)|n\rangle$ corresponding to the parameter λ are degenerate with $|n\rangle$.

We shall discuss discrete symmetries in the following two sections.

11.2   Space Reflection or Parity Operation

Space reflection or *parity* operation is reflection through the origin of the coordinate system. Corresponding to such an operation, there is a unitary operator say $\hat{\mathcal{P}}$, also called the *parity operator*. It is such that for a single particle (spatial) wavefunction $\psi(\mathbf{r})$

$$\hat{\mathcal{P}}\psi(\mathbf{r}) = \psi(-\mathbf{r}), \tag{11.7}$$

and for several particles

$$\hat{\mathcal{P}}\psi(\mathbf{r}_1, \mathbf{r}_2, \cdots, \mathbf{r}_N) = \psi(-\mathbf{r}_1, -\mathbf{r}_2, \cdots, -\mathbf{r}_N). \tag{11.8}$$

The parity operator is Hermitian ($\hat{\mathcal{P}}^\dagger = \hat{\mathcal{P}}$), since for any two wavefunctions $\psi(\mathbf{r})$ and $\phi(\mathbf{r})$ we have

$$
\begin{aligned}
\int \phi^*(\mathbf{r})\,\hat{\mathcal{P}}\psi(\mathbf{r})\,d\mathbf{r} &= \int \phi^*(\mathbf{r})\,\psi(-\mathbf{r})\,d\mathbf{r} \\
&\overset{\mathbf{r}\to-\mathbf{r}}{=} \int \phi^*(-\mathbf{r})\,\psi(\mathbf{r})\,d\mathbf{r} \\
&= \int \left[\hat{\mathcal{P}}\phi(\mathbf{r})\right]^*\psi(\mathbf{r})\,d\mathbf{r} \tag{11.9}
\end{aligned}
$$

which can be generalized to wavefunctions of several particles. It follows from Eq. (11.7) that

$$\mathcal{P}^2 = \hat{I}, \tag{11.10}$$

so that the eigenvalues of $\hat{\mathcal{P}}$ are $+1$ or -1, and the eigenstates are said to be *even* or *odd* respectively. Thus, if $\psi_+(\mathbf{r})$ is an even eigenstate of $\hat{\mathcal{P}}$ and $\psi_-(\mathbf{r})$ is an odd eigenstate, we have

$$\hat{\mathcal{P}}\psi_+(\mathbf{r}) \equiv \psi_+(-\mathbf{r}) = +\psi_+(\mathbf{r}), \quad \text{and} \tag{11.11}$$
$$\hat{\mathcal{P}}\psi_-(\mathbf{r}) \equiv \psi_-(-\mathbf{r}) = -\psi_-(\mathbf{r}). \tag{11.12}$$

We note that

$$\int \psi_+^*(\mathbf{r})\,\psi_-(\mathbf{r})\,d\mathbf{r} \stackrel{\mathbf{r}\to-\mathbf{r}}{=} \int \psi_+^*(-\mathbf{r})\,\psi_-(-\mathbf{r})\,d\mathbf{r} \tag{11.13}$$

$$= -\int \psi_+^*(\mathbf{r})\,\psi_-(\mathbf{r})\,d\mathbf{r}, \tag{11.14}$$

so $\quad \displaystyle\int \psi_+^*(\mathbf{r})\,\psi_-(\mathbf{r})\,d\mathbf{r} = 0. \tag{11.15}$

Thus the eigenstates $\psi_+(\mathbf{r})$ and $\psi_-(\mathbf{r})$ are orthogonal, in accordance with the fact that they belong to different eigenvalues of $\hat{\mathcal{P}}$. They also form a complete set, since any function can be written as

$$\psi(\mathbf{r}) = [\psi_+(\mathbf{r}) + \psi_-(\mathbf{r})], \quad \text{where} \tag{11.16}$$

$$\psi_+(\mathbf{r}) = \frac{1}{2}[\psi(\mathbf{r}) + \psi(-\mathbf{r})], \quad \text{and} \tag{11.17}$$

$$\psi_-(\mathbf{r}) = \frac{1}{2}[\psi(\mathbf{r}) - \psi(-\mathbf{r})], \tag{11.18}$$

Clearly $\psi_+(\mathbf{r})$ has even parity and $\psi_-(\mathbf{r})$ has odd parity.

The action of the Parity operator $\hat{\mathcal{P}}$ on the observables $\hat{\mathbf{r}}$ and momentum $\hat{\mathbf{p}}$ is given by

$$\hat{\mathcal{P}}\hat{\mathbf{r}}\hat{\mathcal{P}}^{-1} = -\hat{\mathbf{r}}, \quad \text{and} \tag{11.19}$$
$$\hat{\mathcal{P}}\hat{\mathbf{p}}\hat{\mathcal{P}}^{-1} = -\hat{\mathbf{p}}, \tag{11.20}$$

We recall that $\hat{\mathcal{P}}^\dagger = \hat{\mathcal{P}} = \hat{\mathcal{P}}^{-1}$.

The parity operation is equivalent to transforming a right-handed coordinate system into a left-handed one. From Eq. (11.10) we know that if the parity operator commutes with the Hamiltonian of the system then parity is conserved and simultaneous eigenstates of Hamiltonian and the parity operator can be formed.

Except for *weak interaction* (which is responsible, for example, for the β-decay of nuclei) parity operator commutes with the Hamiltonians of atomic and nuclear systems

$$\left[\hat{\mathcal{P}}, \hat{H}\right] = 0, \tag{11.21}$$

and parity is conserved.

PARITY OF SPHERICAL HARMONICS

Since $\hat{\mathbf{L}}$ and $\hat{\mathcal{P}}$ commute, the eigenfunctions of $\hat{L}^2$ and $\hat{L}_z$ are also eigenfunctions of $\hat{\mathcal{P}}$. The coordinate representation of the eigenfunction of $\hat{L}^2$ and $\hat{L}_z$ of a particle is given by

$$\psi_{\alpha,l,m}(\mathbf{r}) = R_\alpha(r) Y_{l,m}(\theta,\phi). \tag{11.22}$$

Since the transformation $\mathbf{r} \longrightarrow -\mathbf{r}$ is equivalent in terms of the spherical polar coordinates (r, θ, ϕ) to the following

$$\begin{pmatrix} r \\ \theta \\ \phi \end{pmatrix} \longrightarrow \begin{pmatrix} r \\ \pi - \theta \\ \pi + \phi \end{pmatrix}, \tag{11.23}$$

it can be shown that

$$Y_{l,m}(\theta,\phi) \longrightarrow (-1)^l Y_{l,m}(\theta,\phi)$$

under reflection and so

$$\hat{\mathcal{P}} Y_{l,m}(\theta,\phi) = (-1)^l Y_{l,m}(\theta,\phi). \tag{11.24}$$

11.3   Time Reversal Symmetry

This discrete symmetry operation was formulated by E. P. Wigner in a seminal paper in 1932. The term 'time reversal' is a misnomer according to Wigner who called it 'reversal of motion'. Before we examine the effect of this transformation in quantum mechanics, we recapitulate how time reversal invariance occurs in classical mechanics. We start from Newton's laws of motion for a mass point

$$m \frac{d^2 \mathbf{r}}{dt^2} = -\boldsymbol{\nabla} V(\mathbf{r}). \tag{11.25}$$

Since Newton's equation is second order in t, we can associate to every solution $\mathbf{r}\,(t)$ of Eq. (11.25) another solution

$$\mathbf{r}'\,(t)\ =\ \mathbf{r}\,(-t)\,. \tag{11.26}$$

If, for example, there is a trajectory as in Fig. (11.1) and we let the particle stop at $t = 0$ and reverse its motion so that the then (at $t = 0$) momentum $\mathbf{p} = \mathbf{p}_0$ is reversed, $\mathbf{p}_0 \to -\mathbf{p}_0$; the particle then traverses backward along the same trajectory, so that one cannot distinguish the two trajectories shown in Fig. (11.1). The above reasoning is true if there is no dissipative force present.

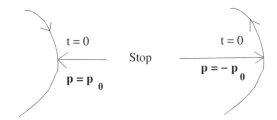

$t = 0$ Stop $t = 0$

$\mathbf{p} = \mathbf{p}_0$ $\mathbf{p} = -\mathbf{p}_0$

Original trajectory   Time reversed trajectory

Figure 11.1: Motion and time reversed motion.

We now consider the Schrödinger equation of wave mechanics

$$i\hbar\frac{\partial \psi\,(\mathbf{r}, t)}{\partial t}\ =\ \left[-\frac{\hbar^2}{2m}\boldsymbol{\nabla}^2 + V\,(r)\right]\psi\,(\mathbf{r}, t)\,. \tag{11.27}$$

If $\psi\,(\mathbf{r}, t)$ is a solution of Eq. (11.27), then $\psi\,(\mathbf{r}, -t)$ is *not* a solution, since the time derivative is of the first order. However $\psi^*\,(\mathbf{r}, -t)$ *is* a solution as can be verified by complex conjugation of Eq. (11.27).

(i). SYMMETRY OPERATIONS IN GENERAL

Before we begin a systematic treatment of the time reversal operation we consider a symmetry transformation given by

$$|\alpha\rangle\ \longrightarrow\ |\tilde{\alpha}\rangle,\qquad |\beta\rangle\ \longrightarrow\ |\tilde{\beta}\rangle. \tag{11.28}$$

For operations like rotation, translation and even parity we require the inner product $\langle\beta|\alpha\rangle$ to remain unchanged:

$$\langle\tilde{\beta}|\tilde{\alpha}\rangle\ =\ \langle\beta|\alpha\rangle \tag{11.29}$$

This is true in the above cases because the symmetry operators are unitary and

$$\langle \tilde{\beta} | \tilde{\alpha} \rangle \;=\; \langle \beta | \hat{U}^{\dagger} \hat{U} | \alpha \rangle \;=\; \langle \beta | \alpha \rangle. \tag{11.30}$$

(ii). <u>ANTI-UNITARY TRANSFORMATION</u>

This condition is too restrictive for time reversal operation (which contains a complex conjugation as well). In this case we have to use the less restrictive and weaker requirement of the transition amplitude

$$|\langle \tilde{\beta} | \tilde{\alpha} \rangle| \;=\; |\langle \beta | \alpha \rangle|. \tag{11.31}$$

The above requiremrnt is, of course, satisfied by unitary transformations. However, the following criterion

$$\langle \tilde{\beta} | \tilde{\alpha} \rangle \;=\; \langle \beta | \alpha \rangle^{*} \;=\; \langle \alpha | \beta \rangle, \tag{11.32}$$

which also satisfies Eq. (11.31) will define *anti-unitary transformations* as we shall now see.

The transformation

$$|\alpha\rangle \;\longrightarrow\; |\tilde{\alpha}\rangle \;=\; \hat{\theta}|\alpha\rangle, \quad |\beta\rangle \;\longrightarrow\; |\tilde{\beta}\rangle \;=\; \hat{\theta}|\beta\rangle, \tag{11.33}$$

is said to be anti-unitary if $\hat{\theta}$ satisfies the following two criteria

(a) $\quad \langle \tilde{\beta} | \tilde{\alpha} \rangle \;=\; \langle \beta | \alpha \rangle^{*}$ $\hspace{3cm}$ (11.34)

(b) $\quad \hat{\theta}\left[C_1 |\alpha\rangle + C_2 |\beta\rangle \right] \;=\; C_1^{*} \hat{\theta}|\alpha\rangle + C_2^{*} \hat{\theta}|\beta\rangle,$ $\hspace{1cm}$ (11.35)

$\quad$ and $\hat{\theta}$ is called an *anti-unitary operator*.

The relation Eq. (11.35) alone defines an *antilinear operator*.

It can be proved that

$$\hat{\theta} \;=\; \hat{U}\hat{K}, \tag{11.36}$$

where $\hat{U}$ is unitary and $\hat{K}$ is the complex conjugation operator. The complex conjugation operator is defined as follows

$$\hat{K}\left[C|\alpha\rangle \right] \;=\; C^{*}\left[\hat{K}|\alpha\rangle \right]. \tag{11.37}$$

Thus

$$|\alpha\rangle = \sum_n |a_n\rangle\langle a_n|\alpha\rangle \xrightarrow{\hat{K}}$$

$$|\tilde{\alpha}\rangle = \sum_n K|a_n\rangle\langle a_n|\alpha\rangle^*$$

$$= \sum_n |a_n\rangle\langle a_n|\alpha\rangle^*, \tag{11.38}$$

where $\hat{K}$ acting on the base ket $|a_n\rangle$ does not change it since the elements of the column matrix

$$|a_n\rangle = \begin{pmatrix} 0 \\ \vdots \\ 0 \\ 1 \\ 0 \\ \vdots \\ 0 \end{pmatrix}$$

are all real.

We now show that $\hat{\theta} = \hat{U}\hat{K}$ is anti-unitary and satisfies Eq. (11.35)

$$\begin{aligned}
\hat{\theta}\left[C_1|\alpha\rangle + C_2|\beta\rangle\right] &= \hat{U}\hat{K}\left[C_1|\alpha\rangle + C_2|\beta\rangle\right] \\
&= \hat{U}\left[C_1^*\hat{K}|\alpha\rangle + C_2^*\hat{K}|\beta\rangle\right] \\
&= C_1^*\hat{U}\hat{K}|\alpha\rangle + C_2^*\hat{U}\hat{K}|\beta\rangle \\
&= C_1^*\hat{\theta}|\alpha\rangle + C_2^*\hat{\theta}|\beta\rangle. \tag{11.39}
\end{aligned}$$

So Eq. (11.35) holds.

$$\begin{aligned}
\text{Now } |\alpha\rangle &\xrightarrow{\hat{\theta}} |\tilde{\alpha}\rangle \\
&= \sum_n \left[\hat{U}\hat{K}|a_n\rangle\right]\langle a_n|\alpha\rangle^* \\
&= \sum_n \left[\hat{U}|a_n\rangle\right]\langle a_n|\alpha\rangle^* \\
&= \sum_n \left[\hat{U}|a_n\rangle\right]\langle\alpha|a_n\rangle, \tag{11.40}
\end{aligned}$$

$$\text{and } |\tilde{\beta}\rangle = \hat{\theta}|\beta\rangle$$
$$= \sum_m \left[\hat{U}|a_m\rangle\right] \langle a_m|\beta\rangle^*.$$

So by dual correspondence

$$\langle \tilde{\beta}| = \sum_m \langle a_m|\beta\rangle \langle a_m|\hat{U}^\dagger, \quad \text{and} \tag{11.41}$$

$$\langle \tilde{\beta}|\tilde{\alpha}\rangle = \sum_m \sum_n \langle a_m|\beta\rangle \langle a_m|\hat{U}^\dagger\hat{U}|a_n\rangle \langle \alpha|a_n\rangle$$
$$= \sum_n \langle \alpha|a_n\rangle \langle a_n|\beta\rangle$$
$$= \langle \alpha|\beta\rangle$$
$$= \langle \beta|\alpha\rangle^*. \tag{11.42}$$

So Eq. (11.34) is satisfied.

We could not evaluate $\langle \tilde{\beta}|$ by considering $\hat{\theta}$ acting on $\langle \beta|$ from the right nor did we define $\hat{\theta}^\dagger$. For that method was valid for linear operators in a ket vector space which was also linear, whereas we are considering an anti-linear operator here.

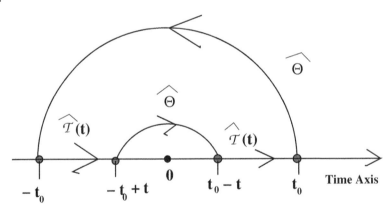

Figure 11.2: Result of successive operations of Time Reversal $\hat{\Theta}$, Time evolution $\hat{\mathcal{T}}(t)$, Time Reversal $\hat{\Theta}$ and Time evolution $\hat{\mathcal{T}}(t)$ on a state at time t_0.

We shall now present a formal theory of time reversal symmetry and associated invariance. We denote time reversal operator by Θ whereas θ denotes any general anti-unitary operator

$$|\alpha\rangle \xrightarrow{\text{time reversal}} |\tilde{\alpha}\rangle = \hat{\Theta}|\alpha\rangle. \tag{11.43}$$

Since $\hat{\Theta}|\alpha\rangle$ is to be a motion reversed state, for a momentum eigenstate $|\mathbf{p}'\rangle$, we expect $\hat{\Theta}|\mathbf{p}'\rangle = |-\mathbf{p}'\rangle$ up to a phase. Likewise for angular momentum eigenstate. We now schematically show the effect of time reversal and time evolution $\mathcal{T}(t)$ in Fig. (11.2). Starting with a state $|\alpha\rangle$ at time t_0, we consider successive operations of time reversal $\hat{\Theta}$ and time evolution $\mathcal{T}(t)$ reaching the point $-t_0 + t$ on the time axis. Then operating by $\hat{\Theta}$ and subsequently operataing by $\mathcal{T}(t)$ we are back to the starting time t_0, where the state is $|\alpha\rangle$. Thus

$$\hat{\mathcal{T}}(t)\hat{\Theta}\hat{\mathcal{T}}(t)\hat{\Theta}|\alpha\rangle = |\alpha\rangle. \tag{11.44}$$

Instead of using Eq. (11.44), we shall consider infinitesimal time evolution of a physical state $|\alpha\rangle$ of a system at time $t = 0$. Then at a slightly later time δt the system will evolve to a state

$$|\alpha, t_0 = 0, t = \delta t\rangle = \left(\hat{I} - i\frac{\hat{H}\delta t}{\hbar}\right)|\alpha\rangle, \tag{11.45}$$

where $\left(\hat{I} - i\frac{\hat{H}\delta t}{\hbar}\right) = \hat{\mathcal{T}}(\delta t)$ is the infinitesimal time evolution operator. Suppose we first apply $\hat{\Theta}$ at $t = 0$ and then let the system evolve in time by $\hat{\mathcal{T}}(\delta t)$, then the state at $t = \delta t$ would be

$$\left(\hat{I} - i\frac{\hat{H}\delta t}{\hbar}\right)\hat{\Theta}|\alpha\rangle.$$

If the motion is symmetric under time reversal we expect that the above state ket would be the same as

$$\hat{\Theta}|\alpha, t_0 = 0, t = -\delta t\rangle,$$

i.e. we first consider a state ket at an earlier time $t = -\delta t$ and then reverse the motion, i.e. reverse $\mathbf{p}$ and $\mathbf{J}$ and we have

$$\left[\hat{I} - i\frac{\hat{H}\delta t}{\hbar}\right]\hat{\Theta}|\alpha\rangle = \hat{\Theta}\left[\hat{I} - i\frac{\hat{H}(-\delta t)}{\hbar}\right]|\alpha\rangle, \quad \text{or} \tag{11.46}$$

$$-i\hat{H}\hat{\Theta}|\alpha\rangle = \hat{\Theta}i\hat{H}|\alpha\rangle. \tag{11.47}$$

Now if $\hat{\Theta}$ were unitary then in Eq. (11.47) we could have calcelled i on both sides [because then $\hat{\Theta} i = i \hat{\Theta}$] and

$$-\hat{H}\hat{\Theta} = \hat{\Theta}\hat{H}. \qquad (11.48)$$

If $|n\rangle$ is an energy eigenket of $\hat{H}$ with eigenvalue E_n then

$$\hat{H}\left[\hat{\Theta}|n\rangle\right] = -\hat{\Theta}\hat{H}|n\rangle = -E_n\left[\hat{\Theta}|n\rangle\right].$$

So $\hat{\Theta}|n\rangle$ would be an energy eigenket of $\hat{H}$ with energy $-E_n$. This does not make any sense, because for a free particle for example the energy spectrum is positive semi-definite from 0 to $+\infty$. There is no state lower than a particle at rest and energy spectrum ranging from $-\infty$ to 0 is physically unacceptable. Free particl Hamiltonian is $\frac{\mathbf{p}^2}{2m}$, $\mathbf{p}$ can change sign but not $\mathbf{p}^2$, Eq. (11.48) would imply

$$\hat{\Theta}^{-1}\frac{\mathbf{p}^2}{2m}\hat{\Theta} = -\frac{\mathbf{p}^2}{2m}.$$

Hence $\hat{\Theta}$ *cannot* be unitary and Eq. (11.48) *cannot* be true. $\hat{\Theta}$ *is anti-unitary* and

$$\hat{\Theta} i \hat{H}|\cdots\rangle = -i\hat{\Theta}\hat{H}|\cdots\rangle$$

for any arbitrary state $|\cdots\rangle$. Then we have from Eq. (11.47)

$$\begin{aligned}\hat{\Theta}\hat{H} &= \hat{H}\hat{\Theta} \\ \left[\hat{H},\hat{\Theta}\right] &= 0. \end{aligned} \qquad (11.49)$$

(iii). <u>MATRIX ELEMENTS $\langle\beta|\hat{\Theta}|\alpha\rangle$</u>

We have indicated that it is best to avoid antiunitary operator acting on the bra vector from the right. However, $\langle\beta|\hat{\Theta}|\alpha\rangle$ is always to be understood as

$$\langle\beta|\hat{\Theta}|\alpha\rangle = \left(\langle\beta|\right)\cdot\left(\hat{\Theta}|\alpha\rangle\right). \qquad (11.50)$$

(iv). OPERATORS UNDER TIME REVERSAL

Often it is convenient to describe operators corresponding to observables which are odd or even under time reversal. We start with an important identity:

$$\langle\beta|\hat{A}|\alpha\rangle \;=\; \langle\tilde{\alpha}|\hat{\Theta}\hat{A}^{\dagger}\hat{\Theta}^{-1}|\tilde{\beta}\rangle, \tag{11.51}$$

where $\hat{A}$ is a linear operator. This identity follows solely from the anti-unitarity of $\hat{\Theta}$.

Proof:

Define

$$|\gamma\rangle \;=\; \hat{A}^{\dagger}|\beta\rangle. \tag{11.52}$$

By dual correspondence

$$|\gamma\rangle \;\overset{\text{D.C.}}{\longleftrightarrow}\; \langle\beta|\hat{A} \;=\; \langle\gamma|, \quad \text{hence} \tag{11.53}$$

$$
\begin{aligned}
\langle\beta|\hat{A}|\alpha\rangle \;&=\; \langle\gamma|\alpha\rangle \\
&=\; \langle\tilde{\alpha}|\tilde{\gamma}\rangle \quad \text{[using Eq. (11.42)]} \\
&=\; \langle\tilde{\alpha}|\hat{\Theta}\hat{A}^{\dagger}|\beta\rangle, \quad \text{or}
\end{aligned}
$$

$$
\begin{aligned}
\langle\beta|\hat{A}|\alpha\rangle \;&=\; \langle\tilde{\alpha}|\left(\hat{\Theta}\hat{A}^{\dagger}\hat{\Theta}^{-1}\right)\hat{\Theta}|\beta\rangle \\
&=\; \langle\tilde{\alpha}|\hat{\Theta}\hat{A}^{\dagger}\hat{\Theta}^{-1}|\tilde{\beta}\rangle, \tag{11.54}
\end{aligned}
$$

which proves the identity.

Now if $\hat{A} = \hat{A}^{\dagger}$, *i.e.* Hermitian then

$$\langle\beta|\hat{A}|\alpha\rangle \;=\; \langle\tilde{\alpha}|\hat{\Theta}\hat{A}\hat{\Theta}^{-1}|\tilde{\beta}\rangle, \quad \text{if} \tag{11.55}$$

$$\hat{\Theta}\hat{A}\hat{\Theta}^{-1} \;=\; \pm\hat{A}, \tag{11.56}$$

we say that the observable $\hat{A}$ is *even* or *odd* under time reversal according to whether we have the *upper* or the *lower* of the signs in Eq. (11.56).

The expectation value of $\hat{A}$ is given by

$$\langle\alpha|\hat{A}|\alpha\rangle \;=\; \pm\langle\tilde{\alpha}|\hat{A}|\tilde{\alpha}\rangle, \tag{11.57}$$

where $\langle\tilde{\alpha}|\hat{A}|\tilde{\alpha}\rangle$ is the expectation value taken with respect to time reversed state. As an example, we consider momentum operator under time reversal.

Since this is motion reversal we expect that momentum operator should be *odd* under time reversal.

$$\hat{\Theta}\hat{\mathbf{p}}\hat{\Theta}^{-1} = -\hat{\mathbf{p}}, \quad \text{so that} \tag{11.58}$$
$$\hat{\mathbf{p}}\hat{\Theta} = -\hat{\Theta}\hat{\mathbf{p}}, \quad \text{and}$$
$$\hat{\mathbf{p}}\left[\hat{\Theta}|\mathbf{p}'\rangle\right] = -\hat{\Theta}\hat{\mathbf{p}}|\mathbf{p}'\rangle$$
$$= -\mathbf{p}'\left[\hat{\Theta}|\mathbf{p}'\rangle\right]. \tag{11.59}$$

Thus $\hat{\Theta}|\mathbf{p}'\rangle$ is momentum eigenket with eigenvalue $-\mathbf{p}'$.

Likewise

$$\hat{\Theta}\hat{\mathbf{x}}\hat{\Theta}^{-1} = \hat{\mathbf{x}} \quad \text{and} \tag{11.60}$$
$$\hat{\mathbf{x}}\left[\hat{\Theta}|\mathbf{x}'\rangle\right] = \hat{\Theta}\hat{\mathbf{x}}|\mathbf{x}'\rangle$$
$$= \mathbf{x}'\left[\hat{\Theta}|\mathbf{x}'\rangle\right]. \tag{11.61}$$

Thus $\hat{\Theta}|\mathbf{x}'\rangle$ is an eigenket of the coordinate operator $\hat{\mathbf{x}}$ with eigenvalue $\mathbf{x}'$.

Using the relations Eq. (11.58) and Eq. (11.60) one can check the invariance of the fundamental commutation relations

$$[\hat{x}_i, \hat{p}_j]|\cdots\rangle = i\hbar\delta_{i,j}|\cdots\rangle$$

under time reversal.

Similarly to preserve

$$\left[\hat{J}_i, \hat{J}_j\right] = i\hbar\,\epsilon_{i,j,k}\,\hat{J}_k,$$

the angular momentum operators must satisfy

$$\hat{\Theta}\hat{\mathbf{J}}\hat{\Theta}^{-1} = -\hat{\mathbf{J}}. \tag{11.62}$$

(v). <u>SPIN HALF PARTICLE</u>

We first construct the eigenket of $\hat{\mathbf{S}} \cdot \hat{\mathbf{n}}$ with the eigenvalue $\frac{\hbar}{2}$. If the polar and the azimuthal angles chracterizing $\hat{\mathbf{n}}$ be β and α respectively, we can rotate the spinor $\begin{pmatrix} 1 \\ 0 \end{pmatrix}$ representing the spin up state first about y axis by angle

Construction of $\sigma \cdot \hat{\mathbf{n}}$ eigenspinor

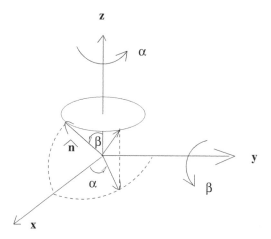

Figure 11.3: Construction of $\boldsymbol{\sigma} \cdot \hat{\mathbf{n}}$ eigenspinor.

β and subsequently rotate by an angle α about z axis and obtain the eigenket of $\hat{\mathbf{S}} \cdot \hat{\mathbf{n}}$ with eigenvalue $\frac{\hbar}{2}$ as indicated in Fig. (11.3). Then

$$|\hat{\mathbf{n}}; +\rangle = \exp\left(-i\frac{\hat{S}_z \alpha}{\hbar}\right) \exp\left(-i\frac{\hat{S}_y \beta}{\hbar}\right) |+\rangle, \quad \text{and} \tag{11.63}$$

$$\hat{\Theta}|\hat{\mathbf{n}}; +\rangle = \hat{\Theta} \exp\left(-i\frac{\hat{S}_z \alpha}{\hbar}\right) \exp\left(-i\frac{\hat{S}_y \beta}{\hbar}\right) |+\rangle$$

$$= \hat{\Theta} \exp\left(-i\frac{\hat{S}_z \alpha}{\hbar}\right) \hat{\Theta}^{-1} \hat{\Theta} \exp\left(-i\frac{\hat{S}_y \beta}{\hbar}\right) \hat{\Theta}^{-1} \hat{\Theta}|+\rangle$$

$$= \exp\left(-i\frac{\hat{S}_z \alpha}{\hbar}\right) \exp\left(-i\frac{\hat{S}_y \beta}{\hbar}\right) \hat{\Theta}|+\rangle$$

$$= \exp\left(-i\frac{\hat{S}_z \alpha}{\hbar}\right) \exp\left(-i\frac{\hat{S}_y \beta}{\hbar}\right) |-\rangle$$

$$\propto |\hat{\mathbf{n}}; -\rangle \tag{11.64}$$

$$= \eta|\hat{\mathbf{n}}; -\rangle \tag{11.65}$$

$$\tag{11.66}$$

To obtain η we obtain $|\hat{\mathbf{n}}; -\rangle$ directly by the Euler rotation of spin up state $|+\rangle$ as follows

$$|\hat{\mathbf{n}}; -\rangle \;=\; \exp\left(-i\frac{\hat{S}_z \alpha}{\hbar}\right) \exp\left(-i\frac{\hat{S}_y\,[\pi + \beta]}{\hbar}\right) |+\rangle. \tag{11.67}$$

From Eq. (11.65) and Eq. (11.67), setting $\hat{\Theta} = \hat{U}\hat{K}$, we get

$$\hat{\Theta} \;=\; \eta \exp\left(-i\frac{\hat{S}_y \pi}{\hbar}\right) \hat{K} \tag{11.68}$$

$$\;=\; -i\eta \left(\frac{2\hat{S}_y}{\hbar}\right) \hat{K}, \tag{11.69}$$

since $\hat{K}$ operating on a base ket does not change it.

Problem 11.1 Expand $\exp\left(-i\frac{\hat{S}_y \pi}{\hbar}\right)$ and obtain Eq. (11.69) for spin $\frac{1}{2}$ particles.

It can then be shown that

$$\hat{\Theta}^2 \;=\; \pm\hat{I}, \tag{11.70}$$

where the upper sign is for spinless or integral spin boson states and the lower sign is for $\frac{1}{2}$ odd integral fermionic states. Hence

$$\hat{\Theta} \;=\; (-1)^{2j}. \tag{11.71}$$

(vi). <u>KRAMER'S DEGENERACY</u>

If the Hamiltonian of a particle is invariant under time reversal then $\hat{\Theta}$ commutes with $\hat{H}$. It then leads to the degeneracy known as *Kramer's degeneracy*. Because then the energy eigenket $|n\rangle$ and its time reversed state $\hat{\Theta}|n\rangle$ belong to the same energy eigenvalue E_n.
Proof:

$$\hat{\Theta}\hat{H}|n\rangle \;=\; \hat{H}\hat{\Theta}|n\rangle \;=\; E_n|n\rangle. \tag{11.72}$$

If $|n\rangle$ and $\hat{\Theta}|n\rangle$ represent the same state then they are non-degenerate differing at most by a phase factor *i.e*

$$
\begin{aligned}
\hat{\Theta}|n\rangle &= e^{i\delta}|n\rangle, \\
\therefore \hat{\Theta}^2|n\rangle &= e^{-i\delta}\hat{\Theta}|n\rangle = e^{-i\delta}e^{+i\delta}|n\rangle \\
&= |n\rangle.
\end{aligned}
\tag{11.73}
$$

But from Eq. (11.71) for $\frac{1}{2}$ odd integral spin this is not possible for which $\hat{\Theta}^2 = -1$.

Thus for $\frac{1}{2}$ odd integral spin particle, the states $|n\rangle$ and $\hat{\Theta}|n\rangle$ are distinct and hence degenerate states. Thus considering electrons in crystals it is observed that odd electron and even electron systems exhibit very different behaviours. Odd electron systems exhibit this degeneracy known as *Kramer's degeneracy* which is a consequence of time reversal invariance as was pointed out by Wigner. We have also seen for spin $\frac{1}{2}$ system (electron) the spin up $|+\rangle$ and its time reversed state $|-\rangle$ do *not* have the same energy in presence of a magnetic field as in Stern-Gerlach experiment. Here Kramer's degeneracy is lifted by application of external magnetic field.

Chapter 12

Approximate Methods

In quantum mechanics as in classical mechanics, there are a relatively few physical problems that can be solved exactly. Thus one has to take recourse to approximate methods. We start with semi-classical approximation in the next section.

12.1 Semiclassical Aprroximation or WKB Method

For particles moving in a sufficiently uniform fields, and for large values of momentum, the equation of motion differs very little from Newton's classical equation. We would study the limiting transition from quantum to classical mechanics which is formally analogous to transtion from wave optics to geometrical optics. This analogy was used in early works that led to the formulation of quantum mechanics.

We represent the wavefunction as

$$\psi\left(\mathbf{r},t\right) \;=\; \exp\left[\frac{i}{\hbar}S\left(\mathbf{r},t\right)\right] \tag{12.1}$$

which when substituted in Schrödinger equation Eq. (4.20) will yield the following equation

$$-\frac{\partial S\left(\mathbf{r},t\right)}{\partial t} \;=\; \frac{\left(\boldsymbol{\nabla}S\left(\mathbf{r},t\right)\right)^{2}}{2m} + V\left(\mathbf{r}\right) - \frac{i\hbar}{2m}\boldsymbol{\nabla}^{2}S\left(\mathbf{r},t\right) \tag{12.2}$$

for a particle of mass m moving in a potential $V\left(\mathbf{r}\right)$.

In the absence of the last term on the right hand side of Eq. (12.2) we have

$$-\frac{\partial S_{0}\left(\mathbf{r},t\right)}{\partial t} \;=\; \frac{\left(\boldsymbol{\nabla}S_{0}\left(\mathbf{r},t\right)\right)^{2}}{2m} + V\left(\mathbf{r}\right) \tag{12.3}$$

This is a partial differential equation of the first order for the real valued *action* function which is defined in terms of the Lagrangian L by the integral

$$S_0\left(\mathbf{r}, t\right) = \int_a^t L\left(\mathbf{r}, \dot{\mathbf{r}}, t'\right) dt'.$$

The trajectory of motion is normal to the sutfaces of constant value of *action*. This is evident from the fact that the momentum of the particle is given by $\mathbf{p} = \boldsymbol{\nabla} S_0\left(\mathbf{r}, t\right)$. By comparing Eq. (12.2) with Eq. (12.3) we see that the transition from the quantum equation of motion to the classical one corresponds formally to the limit $\hbar \to 0$. Since $\hbar$ is a constant quantity, such a limiting process is justified when the terms containing $\hbar$ in Eq. (12.2) are small in comparison with the other terms in the equation. We investigate stationary states for simplicity's sake.

$$\psi\left(\mathbf{r}, t\right) = u\left(\mathbf{r}\right) \exp\left(-i \frac{Et}{\hbar}\right).$$

We can then separate out the explicit time dependence of $S\left(\mathbf{r}, t\right)$ as follows

$$S\left(\mathbf{r}, t\right) = \sigma\left(\mathbf{r}\right) - Et. \tag{12.4}$$

Hence Eq. (12.2) becomes

$$\frac{\left(\boldsymbol{\nabla}\sigma\left(\mathbf{r}, t\right)\right)^2}{2m} + V\left(\mathbf{r}\right) - E - i\hbar \frac{\boldsymbol{\nabla}^2 \sigma\left(\mathbf{r}\right)}{2m} = 0. \tag{12.5}$$

Then neglecting $\hbar$ term we get the classical equation

$$\frac{\left(\boldsymbol{\nabla}\sigma_0\left(\mathbf{r}, t\right)\right)^2}{2m} + V\left(\mathbf{r}\right) - E = 0. \tag{12.6}$$

for the function $\sigma_0\left(\mathbf{r}\right)$, which is related to the classical momentum

$$\mathbf{p} = \boldsymbol{\nabla}\sigma\left(\mathbf{r}\right). \tag{12.7}$$

This is possible if

$$\left(\boldsymbol{\nabla}\sigma_0\left(\mathbf{r}\right)\right)^2 \gg \hbar |\boldsymbol{\nabla}^2 \sigma_0\left(\mathbf{r}\right)|. \tag{12.8}$$

Thus the inequality Eq. (12.8) can be regarded as the condition under which quantum mechanics goes over to classical mechanics.

Using Eq. (12.7), Eq. (12.8) gives the following inequality

$$\mathbf{p}^2 \gg \hbar |\boldsymbol{\nabla} \cdot \mathbf{p}|. \tag{12.9}$$

For one dimensional motion the inequality Eq. (12.9) becomes

$$1 \quad \gg \quad \frac{\hbar \left(dp/dx \right)}{p^2} = \frac{1}{2\pi} \frac{\partial \lambda}{\partial x}, \quad \text{or} \tag{12.10}$$

$$\lambda \quad \gg \quad \frac{\lambda}{2\pi} \frac{\partial \lambda}{\partial x}. \tag{12.11}$$

In other words the variation of the wavelength over the distance $\frac{\lambda}{2\pi}$ must be much smaller than the wavelength itself. If the characteristic size of the system is denoted by a, then $\frac{d\lambda}{dx} \sim \frac{\lambda}{a}$ and Eq. (12.10) becomes

$$\lambda \ll a.$$

We can also express Eq. (12.10) as follows

$$p^3 \quad \gg \quad m\hbar \left| \frac{dV\left(x\right)}{dx} \right|, \quad \text{since} \tag{12.12}$$

$$p^2 \quad = \quad 2m \left(E - V\left(x\right) \right).$$

Thus we conclude from Eq. (12.12) that the classical description of a quantum mechanical system is approximately justified in the case of the motion of a particle with a large momentum in a potential with small gradient.

If Eq. (12.12) is fulfilled we can develop an approximate method of solving quantum mechanical problems based on the introduction of corrections to the classical description. This method is also known as *WKB* approximation after the names of Wentzel, Kramers and Brillouin who first used this method. We would now describe this method.

WKB APPROXIMATION

This Approximation consists of a method of solving Eq. (12.5) for $\sigma\left(\mathbf{r}\right)$ which determines the stationary state wavefunction

$$u\left(\mathbf{r}\right) \quad = \quad \exp \left[\frac{i}{\hbar} \sigma\left(\mathbf{r}\right) \right]. \tag{12.13}$$

The solution of Eq. (12.5) is written as an expansion in $\hbar$ as follows

$$\sigma\left(\mathbf{r}\right) \quad = \quad \sigma_0\left(\mathbf{r}\right) + \left(\frac{\hbar}{i} \right) \sigma_1\left(\mathbf{r}\right) + \left(\frac{\hbar}{i} \right)^2 \sigma_2\left(\mathbf{r}\right) \quad + \quad \cdots. \tag{12.14}$$

If the conditions Eq. (12.9) for semiclassical approximation are satisfied, then the successive terms in the above series are much smaller than the preceding ones, and we can make use of the method of successive approximations in soving Eq. (12.5). Thus substituting Eq. (12.14) in Eq. (12.5) and equating the coefficients of like powers of $\hbar$, we get the following set of coupled equations.

$$\left.\begin{array}{rcl} (\nabla\sigma_0)^2 + 2m\left[V(\mathbf{r}) - E\right] & = & 0 \\ (\nabla\sigma_1)\cdot(\nabla\sigma_0) + \frac{1}{2}\nabla^2\sigma_0 & = & 0 \\ (\nabla\sigma_1)^2 + 2(\nabla\sigma_0)\cdot(\nabla\sigma_2) + \nabla^2\sigma_1 & = & 0 \\ \vdots & & \vdots\ \ \vdots \end{array}\right\} \tag{12.15}$$

Then by solving $\sigma_0(\mathbf{r})$ from the first equation and putting in the second equation we can solve for $\sigma_1(\mathbf{r})$ and so on. Usually only $\sigma_0(\mathbf{r})$ and $\sigma_1(\mathbf{r})$ are calculated. We shall illustrate the method for one dimensional case. Then Eq.(12.15) can be written as

$$(\sigma_0')^2 = p^2, \quad 2\sigma_1' = -\frac{\sigma_0''}{\sigma_0'}, \quad 2\sigma_2' = \sigma_1'' + (\sigma_1')^2, \tag{12.16}$$

where $p^2 = 2m\left[E - V(x)\right]$ and prime denotes derivatives with respect to x. Then σ_1', σ_2' are obtained from the zeroth order

$$\sigma_0' = \pm p(x) = \pm\sqrt{2m\left[E - V(x)\right]} \tag{12.17}$$

by simple differentiation. In particular we have from the second equation of Eq. (12.16)

$$\sigma_1 = -\ln\sqrt{p} + \ln C. \tag{12.18}$$

By integrating Eq. (12.17) we can determine σ_0, and taking into account Eq. (12.18), Eq. (12.14) and Eq. (12.13), we get the WKB wave function up to terms of the order of $\hbar^2$ in the form

$$u(x) = \frac{C}{[E - V(x)]^{\frac{1}{4}}} \exp\left[\frac{i}{\hbar}\int_a^x \sqrt{2m\left[E - V(x')\right]}dx'\right] +$$
$$\frac{C_1}{[E - V(x)]^{\frac{1}{4}}} \exp\left[-\frac{i}{\hbar}\int_a^x \sqrt{2m\left[E - V(x')\right]}dx'\right]. \tag{12.19}$$

The region in which $E > V(x)$ is called *classically permissible region* where

$$k(x) = \frac{1}{\hbar}\sqrt{2m\left[E - V(x)\right]} \tag{12.19}$$

is real and $k\hbar$ is the classical momentum of the particle as a function of the coordinates. In this region the wavefunction Eq. (12.19) can always be written in the following form

$$u(x) = \frac{A}{\sqrt{p}}\cos\left[\int_a^x k(x')\,dx' + \alpha\right], \tag{12.20}$$

where A and α are real constants. The amplitude of $u(x)$ is proportional to the speed of the classical particle. The *turning points* x_i are the roots of the equation $E = V(x)$, where the classical particle comes to rest, *i.e* $p(x_i) = 0$, after that it moves in the opposite direction. The wavefunction Eq. (12.19) becomes infinity at the turning points. Thus this approximation is not valid for small values of momentum of the classical particle. Supposing x_0 is a turning point, we can determine the distance $|x - x_0|$ for which this quasi-classical approximation may still be used. Expanding about $x = x_0$ we have

$$p^2 = 2m[E - V(x)] \approx 2m\left|\left(\frac{dV}{dx}\right)\right||x - x_0|.$$

Substituting in Eq. (12.12) we have

$$|x - x_0| \gg \frac{1}{2}\left[\frac{\hbar^2}{m\left|\frac{dV}{dx}\right|}\right]^{\frac{1}{3}} \quad \text{or} \tag{12.21}$$

$$|x - x_0| \gg \frac{\hbar}{2p} = \frac{\lambda}{4\pi}, \tag{12.22}$$

where λ is the wavelength corresponding to the momentum value at x.

The region $E < V(x)$ is called the *classically forbidden area*, since $k(x)$ becomes imaginary. Writing

$$k(x) = i\chi(x),$$

where $\chi(x) = \sqrt{2m[V(x) - E]}$ is a real valued function and we can write Eq. (12.19) in the form

$$u(x) = \frac{C_1}{[V(x) - E]^{\frac{1}{4}}} \exp\left[\frac{1}{\hbar}\int^x \sqrt{2m[V(x') - E]}dx'\right] +$$
$$\frac{C}{[V(x) - E]^{\frac{1}{4}}} \exp\left[-\frac{1}{\hbar}\int^x \sqrt{2m[V(x') - E]}dx'\right]. \tag{12.23}$$

Neither Eq. (12.19) nor Eq. (12.23) is valid near the classical turning points brcause $u(x)$ becomes infinity at that point. Since valid wavefunction is to be continuous and smooth for all x, the connection formulae should be obtained which can join the two types of wavefunctions across the turning point. The standard procedure is:

(i). Make a linear approximation to $V(x)$ near the turning point x_i.

(ii). Solve the resulting differential equation exactly.

(iii). Match the solution the other two solutions by choosing the various constants of integration appropriately.

Since the calculation is rather technical and cumbersome we do not reproduce it here. We however present the results of such an analysis for a potential well shown in Fig. (12.1) from the standard literature. In Fig. (12.1) is shown schematically a potential having two turning points x_1 and x_2. The wavefunction must behave like Eq. (12.19) in region II and like Eq. (12.23) in regions I and III. The connection formulae for region I and region II can be shown to be achieved by choosing the integration constants in such a way that

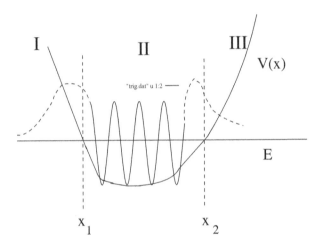

Figure 12.1: WKB wavefunction.

$$\frac{1}{[V(x) - E]^{\frac{1}{4}}} \exp\left[-\frac{1}{\hbar} \int_x^{x_1} \sqrt{2m[V(x') - E]}dx'\right] \longrightarrow$$

$$\frac{2}{[E - V(x)]^{\frac{1}{4}}} \cos\left[\frac{1}{\hbar} \int_{x_1}^x \sqrt{2m[E - V(x')]}dx' - \frac{\pi}{4}\right] \tag{12.24}$$

Similarly from region III to region II we have

$$\frac{1}{[V(x) - E]^{\frac{1}{4}}} \exp\left[-\frac{1}{\hbar} \int_{x_2}^x \sqrt{2m[V(x') - E]}dx'\right] \longrightarrow$$

$$\frac{2}{[E - V(x)]^{\frac{1}{4}}} \cos\left[-\frac{1}{\hbar} \int_x^{x_2} \sqrt{2m[E - V(x')]}dx' + \frac{\pi}{4}\right] \tag{12.25}$$

Since the wavefunction in region II must be unique the arguments of the cosine function in Eq. (12.24) and Eq. (12.25) must differ at most by an integer multiple of π [*not* 2π, because the signs of both sides of Eq. (12.25) can be reversed]. We thus get a consistency condition

$$\int_{x_1}^{x_2} \sqrt{2m\left[E - V\left(x\right)\right]}dx = \left(n + \frac{1}{2}\right)\pi\hbar, \quad n = 0, \ 1, \ 2, \cdots . \tag{12.26}$$

If we ignore the factor $\frac{1}{2}$, then this is simply the quantum condition of old quantum theory of A. Sommerfeld and W. Wilson written in 1915.

$$\oint p\,dq = nh, \tag{12.27}$$

where h is *Planck's* constant h and *not Dirac's* $\hbar$, and the integral is evaluated over one whole period of classical motion from x_1 to x_2 and back.

12.2 Rayleigh Schrödinger Perturbation

This cencerns with finding the modifications in the discrete energy levels and eigenfunctions of a system when a small disturbance (*perturbation*) is applied . It is assumed that the time independent Hamiltonian can be written as

$$\hat{H} = \hat{H}_0 + \hat{H}', \tag{12.28}$$

where the unperturbed Hamiltonian $\hat{H}_0$ is simple in character and the corresponding time-independent Schrödinger equation

$$\hat{H}_0 \left|n^{(0)}\right\rangle = E_n^{(0)}\left|n^{(0)}\right\rangle \tag{12.29}$$

can be exactly solved. The other part $\hat{H}'$ is small enough to be regarded as a perturbation of $\hat{H}_0$. We assume that the set of eigenfunctions $\left\{\left|k^{(0)}\right\rangle\right\}$ corresponding to the set of eigenvalues $\left\{E_k^{(0)}\right\}$ of $\hat{H}_0$ form a complete orthonormal set such that

$$\left\langle k^{(0)}|m^{(0)}\right\rangle = \delta_{k,m}. \tag{12.30}$$

$\left|k^{(0)}\right\rangle$ or $\left|m^{(0)}\right\rangle$ may be discrete or continuous accordingly $\delta_{k,m}$ is extended to include this possibility (i.e. then $\delta\left(k - m\right)$).

The eigenvalue problem we have to solve is

$$\hat{H}|n\rangle = E_n|n\rangle. \tag{12.31}$$

(i)   Non-Degenerate Case

The assumption that $\hat{H}'$ is small suggests that we expand the perturbed eigenfunctions $|n\rangle$ and eigenvalues E_n as a power series in $\hat{H}'$. To this end we introduce a real parameter λ, replace $\hat{H}'$ by $\lambda\hat{H}'$ and express $|n\rangle$ and E_n as a power series in λ in Eq. (12.31). We assume that these two series are continuous and analytic functions of λ, whose value lies between zero and one.

$\lambda = 0 \implies$ corresponds to the unpertubed case,

$\lambda = 1 \implies$ corresponds to full pertubed case,

The particular unperturbed energy level $E_n^{(0)}$ is assumed to be non-degenerate with eigenfunction $|n^{(0)}\rangle$, though other unperturbed energy states may be degenerate. Thus we have

$$\lim_{\lambda \to 0} E_n = E_n^{(0)} \text{ and} \tag{12.32}$$

$$\lim_{\lambda \to 0} |n\rangle = |n^{(0)}\rangle. \tag{12.33}$$

We thus write

$$|n\rangle = |n^{(0)}\rangle + \lambda|n^{(1)}\rangle + \lambda^2|n^{(2)}\rangle + \cdots, \tag{12.34}$$

$$E_n = E_n^{(0)} + \lambda E_n^{(1)} + \lambda^2 E_n^{(2)} + \cdots, \tag{12.35}$$

From Eq. (12.31) we get

$$\left(\hat{H}_0 + \lambda\hat{H}'\right)\left(|n^{(0)}\rangle + \lambda|n^{(1)}\rangle + \lambda^2|n^{(2)}\rangle + \cdots\right) =$$
$$\left(E_n^{(0)} + \lambda E_n^{(1)} + \lambda^2 E_n^{(2)} + \cdots\right) \times$$
$$\left(|n^{(0)}\rangle + \lambda|n^{(1)}\rangle + \lambda^2|n^{(2)}\rangle + \cdots\right) . \tag{12.36}$$

Since the expansion Eq. (12.36) is assumed to be valid for a continuous range of λ, we can equate the coefficients of different powers of λ on both sides and obtain a series of equations that represent successive orders of perturbation.

$$\left(\hat{H}_0 - E_n^{(0)}\right)|n^{(0)}\rangle = 0, \tag{12.37}$$

$$\left(\hat{H}_0 - E_n^{(0)}\right)|n^{(1)}\rangle = \left(E_n^{(1)} - \hat{H}'\right)|n^{(0)}\rangle, \tag{12.38}$$

$$\left(\hat{H}_0 - E_n^{(0)}\right)|n^{(2)}\rangle = \left(E_n^{(1)} - \hat{H}'\right)|n^{(1)}\rangle + E_n^{(2)}|n^{(0)}\rangle, \tag{12.39}$$

$$\cdots \quad \cdots \quad \cdots$$

$$\left(\hat{H}_0 - E_n^{(0)}\right)|n^{(s)}\rangle = \left(E_n^{(1)} - \hat{H}'\right)|n^{(s-1)}\rangle + E_n^{(2)}|n^{(s-2)}\rangle +$$
$$\cdots + E_n^{(s)}|n^{(0)}\rangle. \tag{12.40}$$

We note that any arbitrary multiple of $|n^{(0)}\rangle$ can be added to any of the kets $|n^{(s)}\rangle$ without affecting the left hand side of the equations and hence without affecting the determination of $|n^{(s)}\rangle$ in terms of the lower order terms. We choose the multiple so that

$$\langle n^{(0)}|n^{(s)}\rangle = 0, \quad \text{for} \quad s > 0. \tag{12.41}$$

Also since $\hat{H}_0$ is Hermitian

$$\langle n^{(0)}|\hat{H}_0|n^{(s)}\rangle = \langle n^{(s)}|\hat{H}_0|n^{(0)}\rangle^* = E_n^{(0)}\langle n^{(0)}|n^{(s)}\rangle$$

$$\langle n^{(0)}|\left(\hat{H}_0 - E_n^{(0)}\right)|n^{(s)}\rangle = 0. \tag{12.42}$$

Thus the inner product of $\langle n^{(0)}|$ and the right hand side of Eq. (12.40) is zero and we have

$$E_n^{(s)} = \langle n^{(0)}|\hat{H}'|n^{(s-1)}\rangle. \tag{12.43}$$

(i). FIRST ORDER PERTURBATION

From Eq. (12.43) the first order energy correction is

$$E_n^{(1)} = \langle n^{(0)}|\hat{H}'|n^{(0)}\rangle. \tag{12.44}$$

To calculate $|n^{(1)}\rangle$ we expand it in terms of the complete set $\{|k^{(0)}\rangle\}$ of the unperturbed Hamiltonian $\hat{H}_0$

$$|n^{(1)}\rangle = \sum_k a_{n,k}^{(1)}|k^{(0)}\rangle. \tag{12.45}$$

Here the sum over k includes integration over continuous states, if any. Substitution of Eq. (12.45) in Eq. (12.40) for $s = 1$ yields

$$\left(\hat{H}_0 - E_n^{(0)}\right)\sum_k a_{n,k}^{(1)}|k^{(0)}\rangle + \left(\hat{H}' - E_n^{(1)}\right)|n^{(0)}\rangle = 0, \tag{12.46}$$

$$\text{or} \qquad \left(E_l^{(0)} - E_n^{(0)}\right)a_{n,l}^{(1)} + \langle l^{(0)}|\hat{H}'|n^{(0)}\rangle = E_n^{(1)}\delta_{l,n}. \tag{12.47}$$

For $n \neq l$

$$a_{n,l} = \frac{H'_{l,n}}{E_n^{(0)} - E_l^{(0)}}, \quad \text{where} \tag{12.48}$$

$$H'_{l,n} = \langle l^{(0)}|\hat{H}'|n^{(0)}\rangle.$$

For $n = l$

$$E_n^{(1)} = H'_{n,n}. \tag{12.49}$$

From Eq. (12.45) and Eq. (12.48) we conclude that the perturbation calculation is valid if and only if

$$\left| \frac{H'_{l,n}}{E_n^{(0)} - E_l^{(0)}} \right| \ll 1, \quad \text{for} \quad l \neq n. \tag{12.50}$$

Also from Eq. (12.41) and Eq. (12.45)

$$a_{n,n}^{(1)} = \langle n^{(0)} | n^{(1)} \rangle = 0. \tag{12.51}$$

(ii). <u>SECOND ORDER PERTURBATION</u>

The second order equation is from Eq. (12.39)

$$\left(\hat{H}_0 - E_n^{(0)} \right) | n^{(2)} \rangle = \left(E_n^{(1)} - \hat{H}' \right) | n^{(1)} \rangle + E_n^{(2)} | n^{(0)} \rangle,$$

From Eq. (12.43)

$$E_n^{(2)} = \langle n^{(0)} | \hat{H}' | n^{(1)} \rangle. \tag{12.52}$$

Making use of Eq. (12.45) and Eq. (12.48)

$$\begin{aligned}
E_n^{(2)} &= \sum_{k \neq n} \frac{\langle n^{(0)} | \hat{H}' | k^{(0)} \rangle \langle k^{(0)} | \hat{H}' | n^{(0)} \rangle}{E_n^{(0)} - E_k^{(0)}} \\
&= \sum_{k \neq n} \frac{\left| \langle n^{(0)} | \hat{H}' | k^{(0)} \rangle \right|^2}{E_n^{(0)} - E_k^{(0)}} = \sum_{k \neq n} \frac{\left| \hat{H}'_{n,k} \right|^2}{E_n^{(0)} - E_k^{(0)}}. \tag{12.53}
\end{aligned}$$

The second order state ket $| n^{(2)} \rangle$ can be obtained as follows

$$| n^{(0)} \rangle = \sum_k a_{n,k}^{(2)} | k^{(0)} \rangle. \tag{12.54}$$

Proceeding as in the first order case and making use of Eq. (12.44) and Eq. (12.48) we can obtain

$$a_{n,l}^{(2)} \left(E_l^{(0)} - E_n^{(0)} \right) + \sum_k H'_{l,k} a_{n,k}^{(1)} - E_n^{(1)} a_{n,l}^{(1)} = E_n^{(2)} \delta_{l,n}. \tag{12.55}$$

For $n \neq l$

$$
\begin{aligned}
a_{n,l}^{(2)} &= \frac{1}{E_n^{(0)} - E_l^{(0)}} \left[\sum_k \hat{H}'_{l,k} a_{n,k}^{(1)} - E_n^{(1)} a_{n,l}^{(1)} \right] \\
&= \frac{1}{E_n^{(0)} - E_l^{(0)}} \left[\sum_{k \neq n} \frac{\hat{H}'_{l,k} \hat{H}'_{k,n}}{E_n^{(0)} - E_k^{(0)}} - \frac{\hat{H}'_{n,l} \hat{H}'_{n,n}}{E_n^{(0)} - E_l^{(0)}} \right] \\
&= \sum_{k \neq n} \frac{\hat{H}'_{l,k} \hat{H}'_{k,n}}{\left(E_n^{(0)} - E_k^{(0)} \right) \left(E_n^{(0)} - E_l^{(0)} \right)} - \frac{\hat{H}'_{n,l} \hat{H}'_{n,n}}{\left(E_n^{(0)} - E_l^{(0)} \right)^2}
\end{aligned}
$$

$$(12.56)$$

Because of our choice of $a_{n,n}^{(s)} = 0$, the perturbed state kets are *not* normalized. From Eq. (12.34) putting $\lambda = 1$ we get up to second order in $\hat{H}'$:

$$
E_n = E_n^{(0)} + \langle n^{(0)} | \hat{H}' | n^{(0)} \rangle + \sum_{k \neq n} \frac{\left| \langle n^{(0)} | \hat{H}' | k^{(0)} \rangle \right|^2}{E_n^{(0)} - E_k^{(0)}},
$$

and

$$
\begin{aligned}
|n\rangle &= |n^{(0)}\rangle + \sum_{l \neq n} |l^{(0)}\rangle \left[\frac{H'_{l,n}}{E_n^{(0)} - E_l^{(0)}} - \frac{\hat{H}'_{n,l} \hat{H}'_{n,n}}{\left(E_n^{(0)} - E_l^{(0)} \right)^2} \right. \\
&\qquad \left. + \sum_{k \neq n} \frac{\hat{H}'_{l,k} \hat{H}'_{k,n}}{\left(E_n^{(0)} - E_k^{(0)} \right) \left(E_n^{(0)} - E_l^{(0)} \right)} \right]
\end{aligned}
$$

$$(12.57)$$

We note that

$$
\langle n | n \rangle = 1 + \sum_{k \neq n} \frac{\left| H'_{k,n} \right|^2}{\left(E_n^{(0)} - E_k^{(0)} \right)^2},
$$

$$(12.58)$$

to the second order in $\hat{H}'$ is *not* normalized. We can, however, normalize the perturbed kets (to a given order in λ) by multiplying $|n\rangle$ by a normalization constant $N(\lambda)$, so that

$$
|N(\lambda)|^2 \langle n | n \rangle = 1.
$$

$$(12.59)$$

EXAMPLE

We consider the case of an one dimensional anharmonic oscillator whose Hamiltonian is

$$\hat{H} \;=\; \frac{p^2}{2m} + \frac{1}{2}kx^2 + ax^3 + bx^4, \tag{12.60}$$

$$\hat{H}_0 \;=\; \frac{p^2}{2m} + \frac{1}{2}kx^2, \tag{12.61}$$

$$\text{and } \hat{H}' \;=\; ax^3 + bx^4. \tag{12.62}$$

We shall assume here that $b > 0$ since otherwise the potewntial energy would tend to $-\infty$ for $x \to \pm\infty$, and the energy spectrum would be continuous and unbounded towards negative as well as positive energies.

1-st order energy correction of the n-th state is

$$
\begin{aligned}
E_n^{(1)} \;&=\; a\left(x^3\right)_{n,n} + b\left(x^4\right)_{n,n} \\
&=\; \int_{-\infty}^{+\infty} \left(ax^3 + bx^4\right) \left|\psi_n^{(0)}(x)\right|^2 \, dx.
\end{aligned}
\tag{12.63}
$$

Since the wavefunction ψ_n has definite parity $\left|\psi_n^{(0)}\right|^2$ is always even function of x and x^3 being odd ax^3 term will contribute zero to the integral

$$E_n^{(1)} \;=\; \int_{-\infty}^{+\infty} bx^4 \left|\psi_n^{(0)}(x)\right|^2 \, dx. \tag{12.64}$$

Using oscillator wavefunction it can be shown that

$$\left(x^4\right)_{n,n} \;=\; \frac{3}{4\alpha^4}\left(2n^2 + 2n + 1\right) \tag{12.65}$$

and the first order energy shift is

$$E_n^{(1)} \;=\; b\left(x^4\right)_{n,n} \;=\; \frac{3b}{4}\left(\frac{\hbar}{m\omega}\right)^2 \left(2n^2 + 2n + 1\right). \tag{12.66}$$

The calculation of second order correction to energy is left as an exercise.

(ii) Perturbation Calculations for Degenerate Energy Levels

In the previous calculations we have assumed the initial unperturbed energy level $E_n^{(0)}$ to be non-degerate, having *one* eigenket $|n^{(0)}\rangle$. Now we consider the case when there are two independent kets say $|n^{(0)}\rangle$ and $|m^{(0)}\rangle$ having the same unperturbed energy $E_n^{(0)}$. The level is doubly degenerate. In this case

$$a_{n,l}^{(1)} = \frac{\langle l^{(0)}|\hat{H}'|n^{(0)}\rangle}{E_n^{(0)} - E_l^{(0)}}$$

cannot be calculated for $l = m$ unless $\langle m^{(0)}|\hat{H}'|n^{(0)}\rangle = 0$. We consider the case when $\langle m^{(0)}|\hat{H}'|n^{(0)}\rangle \neq 0$. The initial kets $|m^{(0)}\rangle$ and $|n^{(0)}\rangle$ are orthogonal to the ket $|k^{(0)}\rangle$ with eigenvalue $E_k^{(0)} \neq E_n^{(0)}$. Although they need not be orthogonal to each other it is always possible to construct their linear combinations that are mutually orthogonal and normalized to unity. We can take any such pair of linear combinations of $|n^{(0)}\rangle$ and $|m^{(0)}\rangle$ as the initial states. If the perturbation removes the degeneracy in some order then for finite λ there will be two perturbed states having different energies. Since the perturbed wavefunction $|n\rangle$ and energy E_n are assumed to be continuous analytic functions of λ as $\lambda \to 0$, each of the two states will approach a definite linear combination of $|n^{(0)}\rangle$ and $|m^{(0)}\rangle$. Out of the infinite numbers of orthonormal pairs of linear combinations, the particular pair can be obtained as follows. Let

$$C_n|n^{(0)}\rangle + C_m|m^{(0)}\rangle$$

be the initial unberturbed state. Putting this in Eq. (12.38) we get

$$\left(\hat{H}_0 - E_n^{(0)}\right)|n^{(1)}\rangle = \left(E_n^{(1)} - \hat{H}'\right)\left[C_n|n^{(0)}\rangle + C_m|m^{(0)}\rangle\right]. \tag{12.67}$$

Taking the inner product of the above equation successively with $\langle n^{(0)}|$ and $\langle m^{(0)}|$ we get two equations which can be written in matrix form as follows

$$\begin{pmatrix} \langle m^{(0)}|\hat{H}'|m^{(0)}\rangle - E_n^{(1)} & \langle m^{(0)}|\hat{H}'|n^{(0)}\rangle \\ \langle n^{(0)}|\hat{H}'|m^{(0)}\rangle & \langle n^{(0)}|\hat{H}'|n^{(0)}\rangle - E_n^{(1)} \end{pmatrix} \begin{pmatrix} C_m \\ C_n \end{pmatrix} = \begin{pmatrix} 0 \\ 0 \end{pmatrix} \tag{12.68}$$

The solutions for non-trivial C_m and C_n of these homogeneous algebraic equations exist *if and only if* the *secular determinant* of their coefficients are is zero, namely

$$\begin{vmatrix} \langle m^{(0)}|\hat{H}'|m^{(0)}\rangle - E_n^{(1)} & \langle m^{(0)}|\hat{H}'|n^{(0)}\rangle \\ \langle n^{(0)}|\hat{H}'|m^{(0)}\rangle & \langle n^{(0)}|\hat{H}'|n^{(0)}\rangle - E_n^{(1)} \end{vmatrix} = 0. \tag{12.69}$$

This is a quadratic equation in $E_n^{(1)}$ having the roots

$$E_n^{(1)} = \frac{1}{2} \left[\langle m^{(0)}|\hat{H}'|m^{(0)}\rangle + \langle n^{(0)}|\hat{H}'|n^{(0)}\rangle \right] \pm$$

$$\frac{1}{2}\sqrt{ \left[\langle m^{(0)}|\hat{H}'|m^{(0)}\rangle - \langle n^{(0)}|\hat{H}'|n^{(0)}\rangle \right]^2 + 4 \left| \langle m^{(0)}|\hat{H}'|n^{(0)}\rangle \right|^2 }.$$

$$(12.70)$$

Both the roots of $E_n^{(1)}$ are real, since the diagonal matrix elements of the Hermitian operator $\hat{H}'$ are so. The roots are equal *if and only if*

$$\langle m^{(0)}|\hat{H}'|m^{(0)}\rangle = \langle n^{(0)}|\hat{H}'|n^{(0)}\rangle, \quad \text{and} \quad \langle m^{(0)}|\hat{H}'|n^{(0)}\rangle = 0. \qquad (12.71)$$

and in that case the degeneracy is not removed. Then the coefficients C_n and C_m cannot be determined in the first order calculation.

If on the other hand neither of the two equations above are stisfied then the two values of $E_n^{(1)}$ calculated from Eq. (12.70) are distinct and each can be used in turn to calculate C_n and C_m from Eq. (12.68). We can thus obtain the desired pair of orthonormal combination of the unperturbed wavefunction $|m^{(0)}\rangle$ and $|n^{(0)}\rangle$.

Now to calculate the first order wavefunctions we consider the equation Eq. (12.38)

$$\left(\hat{H}_0 - E_n^{(0)} \right) \sum_{l \neq n,m} a_{n,l}^{(1)}|l^{(0)}\rangle + \left(\hat{H}' - E_n^{(1)} \right) \left(C_n|n^{(0)}\rangle + C_m|m^{(0)}\rangle \right) = 0, \qquad (12.72)$$

and we obtain

$$a_{n,k}^{(1)} \left(E_k^{(0)} - E_n^{(0)} \right) = - \langle k^{(0)}|\hat{H}'|m^{(0)}\rangle C_m - \langle k^{(0)}|\hat{H}'|n^{(0)}\rangle C_n. \qquad (12.73)$$

This gives $a_{n,k}^{(1)}$ for $k \neq n, m$ and also we have from Eq. (12.42) for $s = 1$

$$a_{n,m}^{(1)} = 0 = a_{n,n}^{(1)}. \qquad (12.74)$$

REMOVAL OF DEGENERACY IN THE SECOND ORDER

If the two values of $E_n^{(1)}$ obtained in the first order calculation are equal, one must go to the second order to check whether the degeracy is removed. For this we proceed in the similar fashion as above and obtain from Eq. (12.39)

$$\sum_{k \neq m,n} \langle m^{(0)}|\hat{H}'|k^{(0)}\rangle a_{n,k}^{(1)} - E_n^{(2)} C_m = 0, \qquad (12.75)$$

$$\sum_{k \neq m,n} \langle n^{(0)}|\hat{H}'|k^{(0)}\rangle a_{n,k}^{(1)} - E_n^{(2)} C_n = 0, \qquad (12.76)$$

Again we have taken $a_{n,n}^{(1)} = 0 = a_{n,m}^{(1)}$.

Substituting $a_{n,k}^{(1)}$ from Eq. (12.74)

$$\left[\sum_{k \neq n,m} \frac{\langle m^{(0)}|\hat{H}'|k^{(0)}\rangle \langle k^{(0)}|\hat{H}'|m^{(0)}\rangle}{E_n^{(0)} - E_k^{(0)}} - E_n^{(2)} \right] C_m \ +$$

$$\sum_{k \neq n,m} \frac{\langle m^{(0)}|\hat{H}'|k^{(0)}\rangle \langle k^{(0)}|\hat{H}'|n^{(0)}\rangle}{E_n^{(0)} - E_k^{(0)}} C_n \ = \ 0, \tag{12.77}$$

$$\sum_{k \neq n,m} \frac{\langle m^{(0)}|\hat{H}'|k^{(0)}\rangle \langle k^{(0)}|\hat{H}'|n^{(0)}\rangle}{E_n^{(0)} - E_k^{(0)}} C_m \ +$$

$$\left[\sum_{k \neq n,m} \frac{\langle n^{(0)}|\hat{H}'|k^{(0)}\rangle \langle k^{(0)}|\hat{H}'|n^{(0)}\rangle}{E_n^{(0)} - E_k^{(0)}} - E_n^{(2)} \right] C_n \ = \ 0. \tag{12.78}$$

Again to obtain the solutions for C_m and C_n we have to equate the secular determinant corresponding to the Eq. (12.77) and Eq. (12.78) to zero which will yield roots for $E_n^{(2)}$ which are of the same general form as Eq. (12.70).

Analogous to Eq. (12.71) we have as the conditions for the roots in the second order being equal the equations

$$\sum_{k \neq n,m} \frac{\left| \langle m^{(0)}|\hat{H}'|k^{(0)}\rangle \right|^2}{E_n^{(0)} - E_k^{(0)}} \ = \ \sum_{k \neq n,m} \frac{\left| \langle n^{(0)}|\hat{H}'|k^{(0)}\rangle \right|^2}{E_n^{(0)} - E_k^{(0)}} \tag{12.79}$$

and $\displaystyle \sum_{k \neq n,m} \frac{\langle m^{(0)}|\hat{H}'|k^{(0)}\rangle \langle k^{(0)}|\hat{H}'|n^{(0)}\rangle}{E_n^{(0)} - E_k^{(0)}} \ = \ 0.$ (12.80)

Thus unless both of Eq. (12.79) and Eq. (12.80) are satisfied, the degeneracy is removed in the scond order.

All the foregoing calculations can be similarly extended and generalized to higher orders and also for the case in which the initial state is more than doubly degenerate.

12.3 The Variational Method

The variational method that we now discuss is very useful for estimating the ground state energy E_0 when unlike the perturbation calculation, we do not have any knowledge of the exact solutions to a problem whose Hamiltonian is sufficiently similar.

In this method an arbitrary trial function $|\phi\rangle$ is used to calculate the expectation value of $\hat{H}$

$$\langle \hat{H} \rangle \ = \ \langle \phi|\hat{H}|\phi\rangle. \tag{12.81}$$

$|\phi\rangle$ can be expressed in terms of the complete set of eigenfunctions $\{|n\rangle\}$ of $\hat{H}$:

$$|\phi\rangle \;=\; \sum_n A_n|n\rangle, \quad \text{where} \quad \hat{H}|n\rangle \;=\; E_n|n\rangle. \tag{12.82}$$

$$\therefore\; \langle\phi|\hat{H}|\phi\rangle \;=\; \sum_n \sum_m A_n^* A_m \langle n|\hat{H}|m\rangle$$

$$\;=\; \sum_n |A_n|^2 E_n. \tag{12.83}$$

Replacing E_n in Eq. (12.83) by the lowest eigenvalue E_0

$$\langle\phi|\hat{H}|\phi\rangle \;=\; \sum_n |A_n|^2 E_n \;\geq\; E_0 \sum_n |A_n|^2 \tag{12.84}$$

If $|\phi\rangle$ is normalized

$$\langle\phi|\phi\rangle \;=\; \sum_n |A_n|^2 \;=\; 1$$

then

$$\langle\hat{H}\rangle \;\geq\; E_0. \tag{12.85}$$

However, if $|\phi\rangle$ is not normalized, then

$$E_0 \;\leq\; \frac{\langle\phi|\hat{H}|\phi\rangle}{\langle\phi|\phi\rangle}. \tag{12.86}$$

In actual application of the method one uses a trial function ϕ that depends on a number of parameters. These parameters are varied until the expectation value is minimum. This will give an upper limit for the ground state energy of the system. The fit of the energy will be closer, the closer is the trial function to the eigenfunction.

GROUND STATE OF HELIUM

Problem 12.1 The Helium atom consists of nucleus of charge $+2e$ surrounded by two electrons. Use a Hamiltonian

$$\hat{H} \;=\; -\frac{\hbar^2}{2m}\left(\boldsymbol{\nabla}_1^2 + \boldsymbol{\nabla}_2^2\right) - 2e^2\left(\frac{1}{r_1} + \frac{1}{r_2}\right) + \frac{e^2}{r_{12}}, \tag{12.87}$$

where r_1 and r_2 are the radial distances of the two electrons from the nucleus, and

$r_{12} = |\mathbf{r}_1 - \mathbf{r}_2|$ is the distance between the two electrons.

Choose a trial function

$$\psi(r_1, r_2) = \frac{Z^3}{\pi a_0^3} \exp\left[-\frac{Z}{a_0}(r_1 + r_2)\right],$$

with Z as the variation parameter, to minimize $\langle \hat{H} \rangle$ and obtain value of Z and the lowest upper limit for the ground state of Helium.

Hints:

(i). The expectation value of each kinetic energy operators can be taken as $\frac{Z^2 e^2}{2a_0}$.

(ii). Each of the potential energy term will yield $-2\frac{Ze^2}{a_0}$.

(iii). You can take the expectation value of the interaction energy term to be $\frac{5Ze^2}{8a_0}$

Chapter 13

Methods for Time Dependent Problems

When the Hamiltonian of the system depends on time, the solutions of the Schrödinger equation are non-stationary. Thus the concept of bound states with discrete energy levels and stationary eigenstates should be modified. There are three ways to make these modifications corresponding to the particular kind of approximations, depending on the nature of time dependence of the Hamiltonian. They are:

 (i). the time dependent perturbation,

 (ii). the harmonic approximation,

 (iii). the adiabatic approximation,

 (iv). the sudden approximation.

   We discuss them in subsequent sections.

13.1   Time Dependent Perturbation

In this case we have to work with the Hamiltonian

$$\hat{H}(t) \;=\; \hat{H}_0 + \hat{H}'(t) \tag{13.1}$$

having a simple part which is independent of time and is exactly solvable. The eigenvalues and eigenfunctions of $\hat{H}_0$

$$\hat{H}_0|n\rangle \;=\; E_n|n\rangle \tag{13.2}$$

are completely known. $\hat{H}'$ is the smaller part which depends on time and is the perturbing Hamiltonian. If initially one of the eigenkets $|i\rangle$ of $\hat{H}_0$ is populated at time $t = 0$, (when $\hat{H}'(t=0) = 0$,) $\hat{H}'(t)$ causes transition between eigenkets as time goes on, so that states other than $|i\rangle$ become populated as this is no longer a stationary problem.

The time-dependent Scrödinger equation is given by

$$i\hbar\frac{\partial}{\partial t}|\alpha, t_0; t\rangle = \hat{H}|\alpha, t_0; t\rangle \tag{13.3}$$

where $|\alpha\rangle$ is the state vector at time t. $|\alpha\rangle$ can be expressed in terms of the complete set $\left\{|n\rangle \exp\left(-i\frac{E_n}{\hbar}t\right)\right\}$ as follows

$$|\alpha\rangle = \sum_n C_n(t)|n\rangle e^{-i\frac{E_n}{\hbar}t} \tag{13.4}$$

with the expansion coefficient C_n depending on time. Substituting of Eq. (13.4) in Eq. (13.3) yields

$$\sum_n i\hbar\frac{dC_n}{dt}|n\rangle e^{-i\frac{E_n}{\hbar}t} + \sum_n C_n E_n|n\rangle e^{-i\frac{E_n}{\hbar}t} = \sum_n C_n\left(\hat{H}_0 + \hat{H}'(t)\right)|n\rangle \times$$
$$e^{-i\frac{E_n}{\hbar}t}. \tag{13.5}$$

Using Eq. (13.2) in the right hand side of Eq. (13.5) and taking inner product of the equation with $|k\rangle$ one gets

$$i\hbar\frac{dC_k(t)}{dt}e^{-i\frac{E_k}{\hbar}t} = \sum_n C_n(t)e^{-i\frac{E_n}{\hbar}t}\langle k|\hat{H}'|n\rangle, \tag{13.6}$$

where the orthonormality of $|n\rangle$ has been used.

Introducing *Bohr's angular frequency*

$$\omega_{k,n} = \frac{E_k - E_n}{\hbar}, \tag{13.7}$$

Eq. (13.6) reduces to the following

$$\frac{dC_k(t)}{dt} = (i\hbar)^{-1}\sum_n\langle k|\hat{H}'(t)|n\rangle C_n(t)\exp(i\omega_{k,n}t). \tag{13.8}$$

The group of Eqs. (13.8) is equivalent to the Schrödinger time dependent Eq. (13.3).

INTERACTION PICTURE

We notice that the time evolution of the coefficients and hence of the system is happening through $\hat{H}'$, the perturbing Hamiltonian. We have discussed time evolution in Table 4.1 of Chapter 4. In the *interaction picture*, the time evolution of quantum system happens through the time evolution of state vector and dynamical variables. We define the interaction picture state ket as follows

$$|\alpha, t_0; t\rangle_I = e^{i\hat{H}_0 t/\hbar}|\alpha, t_0; t\rangle_S, \tag{13.9}$$

which coincides with $|\alpha, t_0; t\rangle_S$ at $t = 0$. For observables represented by operators we define

$$\hat{A}_I(t) = e^{i\hat{H}_0 t/\hbar}\hat{A}_S e^{-i\hat{H}_0 t/\hbar}. \tag{13.10}$$

Thus

$$\hat{H}'_I(t) = e^{i\hat{H}_0 t/\hbar}\hat{H}'(t)e^{-i\hat{H}_0 t/\hbar}, \tag{13.11}$$

where $\hat{H}'(t)$ is the perturbing Hamiltonian in the Schrödinger picture. We already know the connection between the Schrödinger and the Heisenberg picture:

$$|\alpha\rangle_H = e^{i\hat{H}t/\hbar}|\alpha, t_0; t\rangle_S \tag{13.12}$$

$$\text{and } \hat{A}_H(t) = e^{i\hat{H}t/\hbar}\hat{A}_S e^{-i\hat{H}t/\hbar}. \tag{13.13}$$

The basic difference between the pair of Eq. (13.9) and Eq. (13.10) defining the interaction picture and the pair Eq. (13.12) and Eq. (13.13) defining the Heisenberg picture is that $\hat{H}_0$ appears in the former and $\hat{H}$ appears in the latter exponentials.

Thus the time evolution of the state ket in interaction picture is given as follows.

$$
\begin{aligned}
i\hbar\frac{\partial}{\partial t}|\alpha, t_0; t\rangle_I &= i\hbar\frac{\partial}{\partial t}\left[e^{i\hat{H}_0 t/\hbar}|\alpha, t_0; t\rangle_S\right]\\
&= -\hat{H}_0 e^{i\hat{H}_0 t/\hbar}|\alpha, t_0; t\rangle_S + e^{i\hat{H}_0 t/\hbar}\left(\hat{H}_0 + \hat{H}'(t)\right)|\alpha, t_0; t\rangle_S\\
&= e^{i\hat{H}_0 t/\hbar}\hat{H}'(t)|\alpha, t_0; t\rangle_S \tag{13.14}\\
&= \left[e^{i\hat{H}_0 t/\hbar}\hat{H}'(t)e^{-i\hat{H}_0 t/\hbar}\right]\left[e^{i\hat{H}_0 t/\hbar}|\alpha, t_0; t\rangle_S\right]\\
&= \hat{H}'_I(t)|\alpha, t_0; t\rangle_I, \tag{13.15}
\end{aligned}
$$

which is a Schrödinger like equation with the total Hamiltonian $\hat{H}$ being replaced by $\hat{H}'_I(t)$.

Also from Eq. (13.10) we get

$$\frac{d}{dt}\hat{A}_I = (i\hbar)^{-1}\left[\hat{A}_I, \hat{H}_0\right], \tag{13.16}$$

where $\hat{A}_I$ does not depend explicitly on time.

Table 13.1 The three types of description of evolution of quantum states.

	Heisenberg Picture	Interaction Picture	Schrödinger Picture
State Vectors	*No change*	*Evolve by $\hat{H}'$*	*Evolve by $\hat{H}$*
Observables	*Evolves by $\hat{H}$*	*Evolve by $\hat{H}_0$*	*No change*

FIRST ORDER PERTURBATION

Now we return to Eq. (13.8), replace $\hat{H}'$ by $\lambda\hat{H}'$ and express the coefficients C_n-s as power series in λ:

$$C_n(t) = C_n^{(0)}(t) + \lambda C_n^{(1)}(t) + \lambda^2 C_n^{(2)}(t) + \cdots . \qquad (13.17)$$

We assume, as in the time-independent case, the series is a continuous analytic function of λ, for λ between zero and 1. From Eq. (13.17) and Eq. (13.8), equating coefficients of corresponding powers of λ and setting $\lambda = 1$, in the final results we obtain

$$\frac{d}{dt}C_k^{(0)}(t) = 0, \quad \frac{d}{dt}C_k^{(s+1)}(t) = (i\hbar)^{-1}\sum_n \langle k|\hat{H}'(t)|n\rangle C_n^{(s)}(t) e^{i\omega_{k,n}t}. \qquad (13.18)$$

These sets of equations can in principle be integrated successively to obtain approximate solutions to any desired order in the perturbation.

The first equation of Eq. (13.18) shows that to the zeroth order all the coeffients $C_k^{(0)}$ are constant in time. These values are the initial conditions of the problem specifying the state of the system before the perturbation is applied. We shall assume that all except *one* of the coefficients $C_k^{(0)}$ are zero, so that the system is in a definite unperturbed energy state. This need not violate the uncertainty principle since the infinite lapse of time before the perturbation is applied makes the determination of the original energy of the system possible with great precision. We can then put

$$C_k^{(0)} = \langle k|m\rangle = \begin{cases} \delta(k-m) & \text{for a continuum energy spectrum} \\ \delta_{k,m} & \text{for a discrete energy spectrum} \end{cases} \qquad (13.19)$$

Then from Eq. (13.18) using Eq. (13.19) we get in the *first order*

$$\frac{d}{dt}C_k^{(1)}(t) = (i\hbar)^{-1}\langle k|\hat{H}'(t)|m\rangle e^{i\omega_{k,m}t}, \qquad (13.20)$$

where $\omega_{k,m} = (E_k - E_m)/\hbar$. The solution is

$$C_m^{(1)}(t) = (i\hbar)^{-1} \int_{-\infty}^{t} \langle m|\hat{H}'(t')|m\rangle dt' \tag{13.21}$$

$$\text{and } C_k^{(1)}(t) = (i\hbar)^{-1} \int_{-\infty}^{t} \langle k|\hat{H}'(t')|m\rangle e^{i\omega_{k,m}t'} dt', \tag{13.22}$$

where we have taken the initial time t_0 in Eq. (13.3) as $-\infty$. The integration constant is so chosen that $C_m^{(1)}(t)$ and $C_k^{(1)}(t)$ vanish at $t = -\infty$.

Thus to the first order in the perturbation, the transition probability corresponding to a transition $m \to k$, namely, the probability that the system initially in the state m be found at time t in the state $k \neq m$ is

$$P_{k,m}^{(1)}(t) = |C_k^{(1)}|^2 = \hbar^{-2} \left| \int_{-\infty}^{t} H_{k,m}'(t') e^{i\omega_{k,m}t'} dt' \right|^2. \tag{13.23}$$

We also have

$$\begin{aligned}
C_m^{(1)}(t) &\approx C_m^{(0)} + C_m^{(1)}(t) \\
&\approx 1 + (i\hbar)^{-1} \int_{t_0}^{t} \langle m|\hat{H}'(t')|m\rangle dt' \\
&\approx \exp\left[-\frac{i}{\hbar} \int_{t_0}^{t} H_{m,m}'(t')\, dt' \right]. \tag{13.24}
\end{aligned}$$

So that $|C_m^{(1)}(t)|^2 \approx 1$ and perturbation principally changes the phase of the initial state.

13.2 Harmonic Perturbation

This is the case when $\hat{H}'(t)$ depends harmonically on time except being turned on at one time and off at a later time. We designate these two times as 0 and t_0 respectively and assume

$$\langle k|\hat{H}'(t')|m\rangle = 2 \langle k|\hat{H}'|m\rangle \sin\omega t'. \tag{13.25}$$

$\langle k|\hat{H}'|m\rangle$ is independent of time and ω is positive. From Eq. (13.25) and Eq. (13.22) we can obtain the first order amplitude at a time $t \geq t_0$:

$$C_k^{(1)}(t \geq t_0) = -\frac{\langle k|\hat{H}'|m\rangle}{i\hbar} \left[\frac{e^{i(\omega_{k,m}+\omega)t_0} - 1}{\omega_{k,m} + \omega} - \frac{e^{i(\omega_{k,m}-\omega)t_0} - 1}{\omega_{k,m} - \omega} \right]. \tag{13.26}$$

The first term is important when $\omega_{k,m} \approx -\omega$ or $E_k \approx E_m - \hbar\omega$ and the second term is important when $\omega_{k,m} \approx \omega$ or $E_k \approx E_m + \hbar\omega$. Thus the first order effect of a perturbation that varies sinusoidally in the time with angular frequency ω is to transfer *to* or *receive* from the system the quantum of energy $\hbar\omega$ on which it acts.

We now consider the situation when the initial state m is a discrete bound state and the final state k is one of the continuum set of dissociated or ionized states so that $E_k > E_m$ and only the second term in Eq. (13.26) needs to be taken into account. The first order transition probability to the state k after the perturbation is removed is given by

$$
\begin{aligned}
P_{k,m}^{(1)}\left(t \geq t_0\right) &= \left|C_k^{(1)}\left(t \geq t_0\right)\right|^2 \\
&= \frac{t_0^2 \left|\langle k|\hat{H}'|m\rangle\right|^2}{\hbar^2} \frac{\sin^2\left[\frac{1}{2}\left(\omega_{k,m} - \omega\right) t_0\right]}{\left[\frac{1}{2}\left(\omega_{k,m} - \omega\right) t_0\right]^2}.
\end{aligned}
\tag{13.27}
$$

The function $\frac{\sin^2 \theta}{\theta^2}$ where $\theta = \frac{1}{2}\left(\omega_{k,m} - \omega\right) t_0$ is plotted in Fig. 13.1 as a function of θ.

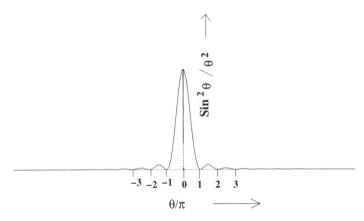

Figure 13.1: Plot of $f\left(\theta\right) = \frac{\sin^2 \theta}{\theta^2}$ where $\theta = \frac{1}{2}\left(\omega_{k,m} - \omega\right) t_0$.

The height of the main peak increases in proportion to t_0^2 and its breadth decreases inversely as t_0, so that the area under the curve is proportional to t_0. Thus if there is a group of states k having energies nearly equal to $E_m + \hbar\omega$ and for which $\langle k|\hat{H}'|m\rangle$ is roughly independent of k then the probability of finding the system in one or another of these states is proportional to t_0.

We are interested to calculate $W_{k,m}$ transition probability per unit time and this implies that a transition has taken place after the perturbation has been on for a time t_0 is proportional to t_0.

(i) Fermi's Golden Rule

The spread of energy of the final state to which transitions occur is connected with energy uncertainty in the following way. $\hat{H}'$ can be looked upon as a device that measures the final energy of the system by transferring it to one of the state k. The time available for measurement is t_0, so that the uncertainty in energy is $\sim \frac{\hbar}{t_0}$. Also the energy conservation is given by $E_k \approx E_m + \hbar\omega$ and suitably modified by uncertainty in principle is a mutual consequence of the calculation and need not be inserted as a separate assumption.

If now $\rho(k)$ is the density of final states such that $\rho(k)\, dE_k$ is the number of final states with energies between E_k and $E_k + dE_k$, then the transition probability per unit time is given by integrating Eq. (13.27) over k and dividing by t_0:

$$W_{k,m} = \frac{1}{t_0} \int \left| C_k^{(1)}(t > t_0) \right|^2 \rho(E_k)\, dE_k. \tag{13.28}$$

Since the breadth of the main peak in Fig. 13.1 becomes small as t_0 becomes large and regarding $\langle k|\hat{H}'|m\rangle$ and $\rho(k)$ as quantities independent of E_k so that they are taken outside the integration sign of Eq. (13.28). Then substituting Eq. (13.27) in Eq. (13.28) and changing the integration variable to $x \equiv \frac{1}{2}(\omega_{k,m} - \omega)t_0$ from E_k and extending the limits of integration to $x = \pm\infty$ to obtain

$$W_{k,m} = \frac{2\pi}{\hbar} \rho(k) \left| \langle k|\hat{H}'|m\rangle \right|^2, \tag{13.29}$$

where we have used the standard integral value

$$\int_{-\infty}^{+\infty} \frac{\sin^2 x}{x^2}\, dx = \pi.$$

This expression of $W_{k,m}$ is independent of t_0 as expected.

The equation Eq. (13.29) is the *Fermi Golden Rule* of probability of transition.

If there are several different groups of final states k_1, k_2, $\cdots$, having about the same energy $E_m + \hbar\omega$, but for which $\langle k_i|\hat{H}'|m\rangle\, \rho(k_i)$ although nearly constant within each group, but differ from one group to another, then Eq. (13.29) with k replaced by k_i, gives the transition probability per unit time t_0 for the ith group.

(ii) Ionization of Hydrogen Atom

We now consider how the first order perturbation theory as discussed above can be applied to calculate the probability of ionization of a hydrogen atom initially in its ground state, when placed in an electric field which varies harmonically with time,

$$\mathcal{E}(t) = 2\mathcal{E}_0 \sin\omega t. \tag{13.30}$$

and the perturbing Hamiltonian is

$$\hat{H}'(t) \;=\; e\boldsymbol{r} \cdot \boldsymbol{\mathcal{E}}(t) \;=\; e\mathcal{E}(t) r \cos\theta, \tag{13.31}$$

where the polar axis and $\boldsymbol{\mathcal{E}}$ are in the direction of positive z axis, and θ is the angle between the position vector $\boldsymbol{r}$ of the electron and $\boldsymbol{\mathcal{E}}$.

The initial ground state of Hydrogen atom is given as follows

$$\langle \boldsymbol{r}|m\rangle \;\equiv\; u_{100}(\boldsymbol{r}) \;=\; \frac{1}{\sqrt{\pi a_0^3}} e^{-\frac{r}{a_0}}. \tag{13.32}$$

The final states should correspond to the motion of a positive energy electron in the Coulomb field of proton which is a scattering state. For simplicity's sake the final states are taken to be free particle plane wave states given by

$$\langle \boldsymbol{r}|\boldsymbol{k}\rangle \;=\; \frac{1}{\sqrt{L^3}} e^{-i\boldsymbol{k}\cdot\boldsymbol{r}}, \tag{13.33}$$

using box normalization, $\hbar\boldsymbol{k}$ being the momentum of the ejected electron.

DENSITY OF FINAL STATES $\rho(\boldsymbol{k})$

The number of states between $\boldsymbol{k}$ and $\boldsymbol{k}+d\boldsymbol{k}$

$$dn \;=\; \frac{L^3}{(2\pi)^3} k^2 dk \sin\theta d\theta d\phi$$

$$dE_k \;=\; d\left(\frac{\hbar^2 k^2}{2\mu}\right) \;=\; \frac{\hbar^2 k dk}{\mu}$$

$$\rho(k) \;=\; \frac{dn}{dE_k} \;=\; \frac{\mu L^3}{8\pi^3\hbar^2} k \sin\theta d\theta d\phi. \tag{13.34}$$

We have expressed the polar angles θ, ϕ of $\boldsymbol{k}$ with respect to some fixed direction which for convenience is taken to be that of the electric field $\boldsymbol{\mathcal{E}}$.

The matrix element is then given by using Eq. (13.25) and Eq. (13.31) and relevant equations

$$\langle \boldsymbol{k}|\hat{H}'|m\rangle \;=\; \frac{e\mathcal{E}_0}{\sqrt{\pi a_0^3 L^3}} \int e^{-ikr\cos\theta'} r \cos\theta'' e^{-\frac{r}{a_0}} d^3\boldsymbol{r}. \tag{13.35}$$

We plot the coordinates and angles for the integration of matrix element in Fig. 13.2, where θ' is the angle between $\boldsymbol{r}$ and $\boldsymbol{k}$ and θ'' is the angle between $\boldsymbol{r}$ and $\boldsymbol{\mathcal{E}}$ while θ denotes the angle between $\boldsymbol{k}$ and the electric field $\boldsymbol{\mathcal{E}}$.

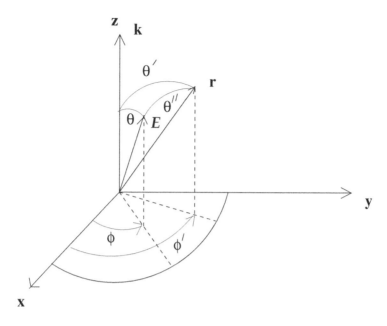

Figure 13.2: Coordinates and angles between $\boldsymbol{k}, \boldsymbol{r}$ and $\boldsymbol{\mathcal{E}}$

Since

$$\cos \theta'' = \cos \theta' \cos \theta + \sin \theta' \sin \theta \cos (\phi' - \phi) \qquad (13.36)$$

$$\text{and } d^3\boldsymbol{r} = r^2 dr \; d(\cos \theta') \, d\phi'. \qquad (13.37)$$

Putting these values in Eq. (13.35) and integrating over ϕ' gives zero for the second term in Eq. (13.36). Then from the first term of Eq. (13.36) one gets

$$\langle k|\hat{H}'|m \rangle = -i \frac{32\pi e \mathcal{E}_0 k a_0^5 \cos \theta}{\sqrt{\pi a_0^3 L^3} \left(1 + k^2 a_0^2\right)^3}. \qquad (13.38)$$

This equation, together with Eq. (13.34) for $\rho(k)$ finally gives the probability per unit time that the electron of the Hydrogen atom from the $1S$ state is ejected into the solid angle $d\Omega = \sin \theta d\theta d\phi$, from the extression Eq. (13.29).

$$W = \frac{256\mu k^3 e^2 \mathcal{E}_0^2 a_0^7}{\pi \hbar^3 \left(1 + k^2 a_0^2\right)^6} \cos^2 \theta \sin \theta d\theta d\phi. \qquad (13.39)$$

13.3 Adiabatic Approximation

This approximation is applicable when the Hamiltonian is slowly varying with time. In this case, we expect that that the solutions of the Schrödinger equation can be approximated by the stationary eigenfunctions of the instaneous Hamiltonian, so that a particular eigenfunction at one time goes over continuously into the corresponding eigenfunction at a later time. Thus if the eigenvalue equation

$$\hat{H}(t)\, u_n(t) \;=\; E_n(t)\, u_n(t) \tag{13.40}$$

can be solved at each instant of time it is expected that that a system that is in a discrete non-degenerate state $u_m(0)$ with energy $E_m(0)$ at $t = 0$ is likely to be in the state $u_m(t)$ with energy $E_m(t)$ at time t, provided that $\hat{H}(t)$ changes very slowly with time. The wave function ψ satisfies the time dependent Schrödinger equation

$$i\hbar \frac{\partial}{\partial t}\psi(t) \;=\; \hat{H}(t)\,\psi(t). \tag{13.41}$$

We expand ψ in terms of the $u_n(t)$s in Eq. (13.40) which form a complete set:

$$\psi(t) \;=\; \sum_n C_n(t)\, u_n(t)\exp\left[(i\hbar)^{-1}\int_0^t E_n(t')\,dt'\right]. \tag{13.42}$$

$u_n(t)$s are assumed to be orthonormal, discrete and non-degenerate. From Eq. (13.42) and Eq. (13.40)

$$\sum_n \left[\frac{dC_n(t)}{dt} u_n(t) + C_n(t)\frac{\partial u_n(t)}{\partial t}\right]\exp\left[(i\hbar)^{-1}\int_0^t E_n(t')\,dt'\right] \;=\; 0, \tag{13.43}$$

where Eq. (13.40) has been used. Then

$$\frac{dC_n(t)}{dl} \;=\; \sum_n C_n(t)\,\langle k|\dot{n}\rangle\exp\left[(i\hbar)^{-1}\int_0^t (E_n(t') - E_k(t'))\,dt'\right] \tag{13.44}$$

$$\text{with } \langle k|\dot{n}\rangle \;\equiv\; \int u_k^* \frac{\partial u_n}{\partial t} d^3\boldsymbol{r}.$$

We also get from Eq. (13.40)

$$\frac{\partial \hat{H}(t)}{\partial t} u_n(t) + \hat{H}(t)\frac{\partial u_n(t)}{\partial t} \;=\; \frac{\partial E_n(t)}{\partial t} u_n(t) + E_n(t)\frac{\partial u_n(t)}{\partial t}.$$

Then

$$\langle k|\frac{\partial \hat{H}(t)}{\partial t}|n\rangle \;=\; (E_n(t) - E_k(t))\,\langle k|\dot{n}\rangle. \tag{13.45}$$

CHOOSING PHASES

Since the u_n-s are normalized

$$\langle n|n\rangle \;=\; 1, \;\;\Longrightarrow\;\; \langle \dot{n}|n\rangle + \langle n|\dot{n}\rangle \;=\; 0.$$

Since the two terms on the left hand side of the second equation are complex conjugate of each other, each is purely imaginary. We can thus write

$$\langle n|\dot{n}\rangle = i\alpha(t),$$

where α is real.

Since the phases of the eigenfunctions are arbitrary at each instant of time we can change the phase of u_n by amount $\gamma(t)$. For the new eigenfunction $u'_n \equiv u_n \exp[i\gamma(t)]$,

$$\langle u'_n|\dot{u}'_n\rangle = \langle u_n|\dot{u}_n\rangle + i\dot{\gamma} = i(\alpha + \dot{\gamma}).$$

Thus choosing

$$\gamma(t) = -\int^t \alpha(t')\,dt'$$

will make $\langle u'_n|\dot{u}'_n\rangle = 0$. We can assume that all phases have been chosen in this way and omit the primes.

Then substituting Eq. (13.45) in Eq. 13.44) we obtain

$$\frac{dC_k(t)}{dt} = = \sum_{n\neq k} \frac{C_n(t)}{\hbar\omega_{k,n}} \left\langle k\left|\frac{\partial \hat{H}}{\partial t}\right|n\right\rangle \exp\left[i\int^t \omega_{k,n}(t')\,dt'\right]. \qquad (13.46)$$

This group of equations Eq. (13.46) is equivalent to the time dependent Schrödinger equation Eq. (13.41).

To estimate the value $\frac{dC_k(t)}{dt}$, we assume that all the quantities $(C_n,\ \omega_{k,n},\ u_n(t),\ \frac{\partial \hat{H}}{\partial t})$ that appear on the right hand sise of Eq. (13.46) which are expected to be slowly varying in time are actually constants in time. I, further, the system is assumed to be in state m at $t = 0$, we can put $C_n(t) \approx \delta n, m$. Thus

$$\frac{dC_k(t)}{dt} \approx (\hbar\omega_{k,m})^{-1} \left\langle k\left|\frac{\partial \hat{H}}{\partial t}\right|m\right\rangle e^{i\omega_{k,m}t}. \quad \text{for} \quad k \neq m,$$

which can readily be integrated and we obtain

$$C_k(t) \approx \left(i\hbar\omega^2_{k,m}\right)^{-1} \left\langle k\left|\frac{\partial \hat{H}}{\partial t}\right|m\right\rangle \left(e^{i\omega_{k,m}t} - 1\right). \quad \text{for} \quad k \neq m. \qquad (13.47)$$

Eq. (13.47) shows that the probability amplitude for a state other than the initial state oscillates in time and does not increase steadily with the long periods of time even though $\hat{H}$ changes by finite amount.

13.4   The Sudden Approximation

In this section we shall consider the case for which the Hamiltonian $\hat{H}$ changes very rapidly. Let us first consider the situation in which the Hamoltonian $\hat{H}$ changes instantaneously at $t = 0$ from $\hat{H}_0$ to $\hat{H}_1$ where both $\hat{H}_0$ and $\hat{H}_1$ are independent of time. Thus for $t < 0$ we have $\hat{H} = \hat{H}_0$ with

$$\hat{H}_0 \psi_k^{(0)} = E_k^{(0)} \psi_k^{(0)}. \tag{13.48}$$

For $t > 0$, $\hat{H} = \hat{H}_1$ and we have

$$\hat{H}_1 \phi_n^{(1)} = E_n^{(1)} \phi_n^{(1)}. \tag{13.49}$$

The subscripts emphasize which Hamiltonians we are dealing with. The sets of eigenfunctions $\left\{ \psi_k^{(0)} \right\}$ abd $\left\{ \phi_n^{(1)} \right\}$ can be assumed to form complete orthonormal sets (which need not be only discrete), so that we can express the general solution of the time dependent Schrödinger equation in terms of them as follows

$$\psi(t) = \sum_k c_k^{(0)} \psi_k^{(0)} \exp \left[-i \frac{E_k^{(0)} t}{\hbar} \right], \quad \text{for } t < 0 \tag{13.50}$$

$$\text{and } \psi(t) = \sum_n d_n^{(1)} \phi_n^{(1)} \exp \left[-i \frac{E_n^{(1)} t}{\hbar} \right], \quad \text{for } t > 0, \tag{13.51}$$

where *summation* implies *sum* over the entire set (discrete plus continuous) of eigenfunctions. The time independent coefficients $c_k^{(0)}$ and $d_n^{(1)}$ for a normalized ψ are respectively the probability amplitude of finding the syatem in the state $\psi_k^{(0)}$ at $t < 0$ and in the state $\phi_n^{(1)}$ at $t > 0$.

Now, since Eq. (13.41) is first order in time, the wavefunction $\psi(t)$ must be a continuous function of t and is thus true particularly at $t = 0$. We therefore equate the two solutions Eq. (13.50) and Eq. (13.51) at $t = 0$ and obtain

$$\sum_k c_k^{(0)} \psi_k^{(0)} = \sum_n d_n^{(1)} \phi_n^{(1)}, \tag{13.52}$$

$$\text{or} \quad d_n^{(1)} = \sum_k c_k^{(0)} \left\langle \phi_n^{(1)} \mid \psi_k^{(0)} \right\rangle. \tag{13.53}$$

This relation is an exact one for the ideal case of an Hamiltonian $\hat{H}$ changing instantaneously at $t = 0$. In practice, however the change in $\hat{H}$ happens during a finite time interval τ. If this time τ is very short then we may in the first approximation set $\tau = 0$ and continue to use Eq. (13.53) to obtain the probability amplitude $d_n^{(1)}$; this procedure is known as the *sudden approximation*.

We can derive a simple criterion of the validity of the sudden approximation as follows. We assume that $\hat{H} = \hat{H}_0$ for $t < 0$, $\hat{H} = \hat{H}_1$ for $t > \tau$ and that during the period $0 < t < \tau$ we have $\hat{H} = \hat{H}_i$ where $\hat{H}_i$ is also time independent. If $\left\{\chi_l^{(i)}\right\}$ denotes a complete, orthonormal set of eigenfunctions of $\hat{H}_i$, such that

$$\hat{H}_i \chi_l^{(i)} = E_l^{(i)} \chi_l^{(i)} \tag{13.54}$$

$$\text{and } \psi(t) = \begin{cases} \sum_k c_k^{(0)} \psi_k^{(0)} e^{-i\frac{E_k^{(0)}t}{\hbar}}, & \text{for } t < 0, \\ \sum_l a_l^{(i)} \chi_l^{(i)} e^{-i\frac{E_l^{(i)}t}{\hbar}} & \text{for } 0 < t < \tau, \\ \sum_n d_n^{(0)} \phi_n^{(1)} e^{-i\frac{E_n^{(1)}t}{\hbar}}, & \text{for } \tau < t, \end{cases} \tag{13.55}$$

where $a_l^{(i)}$-s are also time independent. Using continuity of wavefunction at $t = 0$ we get

$$\sum_k c_k^{(0)} \psi_k^{(0)} = \sum_l a_l^{(i)} \chi_l^{(i)} \tag{13.56}$$

$$\text{or} \quad a_l^{(i)} = \sum_k c_k^{(0)} \left\langle \chi_l^{(i)} \mid \psi_k^{(0)} \right\rangle. \tag{13.57}$$

Similarly, the continuity of $\psi(t)$ at $t = \tau$ yields

$$\sum_l a_l^{(i)} \chi_l^{(i)} e^{-i\frac{E_l^{(i)}\tau}{\hbar}} = \sum_n d_n^{(1)} \phi_n^{(1)} e^{-i\frac{E_n^{(1)}\tau}{\hbar}}. \tag{13.58}$$

From which we obtain

$$d_n^{(1)} = \sum_l a_l^{(i)} \left\langle \phi_n^{(1)} \mid \chi_l^{(i)} \right\rangle e^{i\frac{\left(E_n^{(1)} - E_l^{(i)}\right)\tau}{\hbar}}, \tag{13.59}$$

and using Eq. (13.57) we can write

$$d_n^{(1)} = \sum_k \sum_l c_k^{(0)} \left\langle \phi_n^{(1)} \mid \chi_l^{(i)} \right\rangle \left\langle \chi_l^{(i)} \mid \psi_k^{(0)} \right\rangle e^{i\frac{\left(E_n^{(1)} - E_l^{(i)}\right)\tau}{\hbar}}. \tag{13.60}$$

This exact relation Eq. (13.60) reverts to sudden approximation Eq. (13.53) if τ is small compared to $\frac{\hbar}{|E_n^{(1)} - E_l^{(i)}|}$.

We now return to Eq. (13.53). If the syatem is initially for $t < 0$ in a particular stationary state $\psi_a^{(0)} \exp\left[-i\frac{E_a^{(0)}t}{\hbar}\right]$ where $\psi_a^{(0)}$ is an eigenstate of $\hat{H}_0$. Then $c_k^{(0)} = \delta_{k,a}$, so that the probability amplitude of finding the system in the eigenstate $\phi_n^{(1)}$ of $\hat{H}_1$ after the sudden change in the Hamiltonian has occurred is

$$d_n^{(1)} = \left\langle \phi_n^{(1)} \mid \psi_a^{(0)} \right\rangle. \tag{13.61}$$

EXAMPLE

We consider a charged linear harmonic oscillator acted upon by a spatially uniform time dependent electric field $\mathcal{E}(t)$. The Hamiltonian is therefore

$$\hat{H}(t) = -\frac{\hbar^2}{2m}\frac{d^2}{dx^2} + \frac{1}{2}kx^2 - q\mathcal{E}(t)x \tag{13.62}$$

$$= -\frac{\hbar^2}{2m}\frac{d^2}{dx^2} + \frac{1}{2}k[x - a(t)]^2 - \frac{1}{2}ka(t)^2, \tag{13.63}$$

$$\text{with } a(t) = \frac{q\mathcal{E}(t)}{k}.$$

At $t = 0$ the electric field is switched on suddenly, i.e in a time τ much shorter than ω^{-1}, where ω is the angular frequency of the oscillator, and afterwards it assumed to have the constant value $\mathcal{E}_0$. Thus

$$\hat{H} = \hat{H}_0 = -\frac{\hbar^2}{2m}\frac{d^2}{dx^2} + \frac{1}{2}kx^2, \quad \text{when } t < 0, \tag{13.64}$$

$$\hat{H} = \hat{H}_1 = -\frac{\hbar^2}{2m}\frac{d^2}{dx^2} + \frac{1}{2}k(x-a)^2 - \frac{1}{2}a^2, \quad \text{when } t > \tau, \tag{13.65}$$

$$\text{where } a = \frac{q\mathcal{E}_0}{k} = \frac{q\mathcal{E}_0}{m\omega^2}. \tag{13.66}$$

The Hamiltonian $\hat{H}_0$ is that of a linear hatmonic oscillator.

Problem 13.1 Assume that the oscillator is initially (at $t \leq 0$) in its ground state $\psi_0(x)$.

(i). Show that $E_n^{(1)} = \left(n + \frac{1}{2}\right)\hbar\omega - \frac{q^2\mathcal{E}_0^2}{2m\omega^2}$, $n = 0, 1, 2, \cdots$.

(ii). Calculate $d_n^{(1)}$ from Eq. (13.61)

Chapter 14

Scattering Theory I

Introduction

In a collision problem, the energy of the incident particle is specified in advance, unlike the bound state case where the boundary conditions on the wavefunctions at large distances give rise to the quantization of energy into discrete energy levels of the system. In the scattering problems the behaviour of the wavefunction at large distances is obtained in terms of the energy of the incident particle. We discuss some of the simpler theories in this chapter, deferring some of the formal theories for the next chapter.

14.1 Scattering Experiments: Cross Section

We shall be concerned primarily with collision in three dimensions in which a particle collides with a fixed force field or two particles collide with each other.

We consider an incident beam of monoenergetic particles from a source which is collimated by a slit so that all the particles in the beam move in the same direction. The beam is scattered by the target and the number of particles scattered per unit time in the direction specified by polar angles (θ, ϕ) is measured. Thus to study the interaction between protons and nuclei of atoms of gold, a proton beam is allowed to interact with a target consisting of of a thin film of scatterer to ensure single collisions. Since quantum mechanics does not allow the concept of a well defined trajectory the angle of scattering cannot be calculated precisely and only the probability, of scattering into a certain direction can be predicted. To count the scattered particles in a collision experiment a detector is placed outside the path of the incident beam. If the detector subtends a solid angle $d\Omega$ at the scattering centre in the direction (θ, ϕ), the number of particles $N d\Omega$ entering the detector per unit time is proportional to the incident flux F, defined as the number of particles per unit time crossing a unit area placed normal to the direction of incidence. Assuming that the beam is uniform and that the density of particles in the beam is uniform and is sufficiently small so as to avoid interference between the beam particles, the *differential scattering cross section* $\frac{d\sigma}{d\Omega}$ can be defined as

$$\frac{d\sigma}{d\Omega} = \frac{N}{F}. \tag{14.1}$$

The *total scattering cross section* is defined as the integral of $\frac{d\sigma}{d\Omega}$ over all solid angles

$$\sigma_{\text{total}} = \int \left(\frac{d\sigma}{d\Omega}\right) d\Omega = \int_0^{2\pi} d\phi \int_0^\pi \left(\frac{d\sigma}{d\Omega}\right) \sin\theta d\theta. \tag{14.2}$$

Since N is the number of particles per unit time, while F is the number of particles per unit area per unit time, the dimensions of $\frac{d\sigma}{d\Omega}$ and of σ_{total} are those of area.

RELATIONS BERWEEN ANGLES IN THE LABORATORY AND THE CENTRE OF MASS SYSTEMS

The relations between the scattering angles in the laboratory system and in the centre of mass system can be obtained from the definitions of the co-ordinate systems translating the laboratory frame in the direction of the incident particle with speed so as to bring the centre of mass to rest. In Fig. 14.1 we show a particle of mass m_1 and initial speed v to the right colliding with a particle of mass m_2 that is

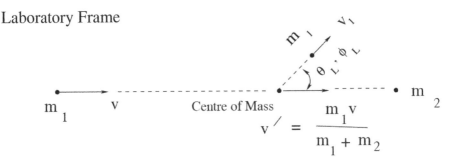

Figure 14.1: Laboratory Frame: in which the incident particle m_1 has velocity $\boldsymbol{v}$, the target particle m_2 is initially at rest, and the centre of mass (c.m) has speed $v' = \frac{m_1 v}{m_1 + m_2}$.

initially at rest in the laboratory frame. The centre of mass then moves to the right with the speed $v' = \frac{m_1 v}{m_1 + m_2}$, according to the law of momentum consevation.

Then in Fig. (14.2) we show that in the the centre of mass system the particles of masses m_1 and m_2 approach the centre of mass with speeds

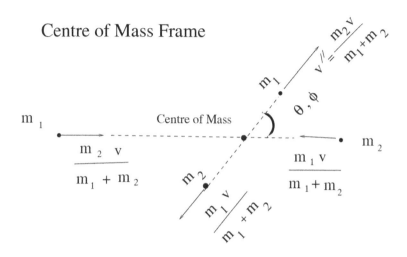

Figure 14.2: Centre of Mass Frame: in which the two particle of masses m_1 m_2 approach the stationary centre of mass with speeds $\frac{m_2 v}{m_1 + m_2}$ and $\frac{m_1 v}{m_1 + m_2}$ respectively. After the collision they recede from the stationary centre of mass again with the same speeds but in an spherical angle (θ, ϕ) from the original line of approach.

$$v'' = v - v' = \frac{m_2 v}{m_1 + m_2} \text{ and } v' = \frac{m_1 v}{m_1 + m_2}$$

respectively. For elastic collision which we take to be the case here, the particles recede from the centre of mass after collision with the same speeds as is depicted in the figure. From the vector diagram in Fig. (14.3) we obtain

$$
\begin{aligned}
v'' \cos\theta + v' &= v_1 \cos\theta_L, \\
v'' \sin\theta &= v_1 \sin\theta_L, \\
\phi &= \phi_L.
\end{aligned}
\tag{14.3}
$$

From the first two of Eq. (14.3) we get

$$\tan\theta_L = \frac{\sin\theta}{\tau + \cos\theta}, \quad \tau = \frac{v'}{v''} = \frac{m_1}{m_2}, \tag{14.4}$$

Laboratory to Centre–of–mass frame transformation

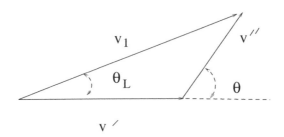

Figure 14.3: Laboratory to the Centre of mass frame transformation: in which we show the vector addition of velocity v' of the canter of mass in the laboratory frame to the the velocity v'' of the observed scattered particle in the centre of mass frame to give the velocity v_1n the laboratry frame.

CROSS SECTIONS

The relation between the scattering cross sections in the laboratory and the centre of mass frames is obtained by equating the number of particles scattered in $d\Omega$ about (θ, ϕ) and in $d\Omega_L$ about (θ_L, ϕ_L):

$$\left(\frac{d\sigma}{d\Omega}\right)_L \sin\theta_L d\theta_L = \left(\frac{d\sigma}{d\omega}\right) \sin\theta d\theta. \tag{14.5}$$

Problem 14.1 Use Eq. (14.3) Eq. (14.4) and Eq. (14.4) to obtain

$$\left(\frac{d\sigma}{d\Omega}\right)_L = \frac{[1 + \tau^2 + 2\tau\cos\theta]^{\frac{3}{2}}}{|1 + \cos\theta|}\left(\frac{d\sigma}{d\Omega}\right). \tag{14.6}$$

14.2   Potential Scattering

We now consider the scattering of a beam of particles by a fixed centre of force. The Schrödinger equation

$$i\hbar\frac{\partial}{\partial t}\psi(\mathbf{r}, t) = \left[-\frac{\hbar^2}{2m}\mathbf{\nabla}^2 + V(\mathbf{r})\right]\psi(\mathbf{r}, t) \tag{14.7}$$

describes the motion, m being the mass of the particle. Since $V(\mathbf{r})$ does not depend on time, we look for stationary solutions

$$\psi(\mathbf{r}, t) = u(\mathbf{r}) \exp\left(-i\frac{Et}{\hbar}\right). \tag{14.8}$$

Here $u(\mathbf{r})$ satisfies

$$\left[-\frac{\hbar^2}{2m}\nabla^2 + V(\mathbf{r})\right] u(\mathbf{r}) = Eu(\mathbf{r}), \tag{14.9}$$

where E, the energy of the incident particle has definite value

$$E = \frac{p^2}{2m} = \frac{1}{2}mv^2, \tag{14.10}$$

$\mathbf{v}$ being the incident velocity and $\mathbf{p}$ the momentum of the incident particle.

We write Eq. (14.9) as

$$\left[\nabla^2 + k^2 - V_{\text{eff}}(\mathbf{r})\right] u(\mathbf{r}) = 0, \tag{14.11}$$
$$\text{where } k = \frac{p}{\hbar}$$

is the wavenumber of the particle and

$$V_{\text{eff}}(\mathbf{r}) = \frac{2m}{\hbar^2}V(\mathbf{r}) \tag{14.12}$$

is the effective potential.

We shall consider short range potential $V_{\text{eff}}(\mathbf{r})$ which either goes to zero beyond a certain distance or decreases exponentially with r, in which case for large r, $V_{\text{eff}}(\mathbf{r})$ can be neglected in Eq. (14.11) and it reduces to the free particle equation

$$\left[\nabla^2 + k^2\right] u(\mathbf{r}) = 0. \tag{14.13}$$

In the asymptotic region $r \to \infty$ $u(r)$ must describe both the incident wave and also the scattered particles, so that

$$\lim_{r\to\infty} u(\mathbf{r}) \longrightarrow u_{\text{inc}}(\mathbf{r}) + u_{\text{sc}}(\mathbf{r}). \tag{14.14}$$

We take the direction of the incident beam as the z-axis and the incident wave can be represented by a plane wave

$$u_{\text{inc}}(\mathbf{r}) = \exp(ikz). \tag{14.15}$$

Since $u_{\text{inc}}(\mathbf{r})$ is normalized to unity, it represents a 'beam' with one particle per unit volume which travels with a velocity $\mathbf{v} = \frac{\mathbf{p}}{m}$ and the incident flux F is

$$F = v. \tag{14.16}$$

Now, far away from the scattering centre the wave function $u_{sc}(\mathbf{r})$ should be a outward moving spherical wave having the form

$$u_{sc}(\mathbf{r}) \; = \; f(k,\theta,\phi)\frac{e^{ikr}}{r}, \tag{14.17}$$

where (r,θ,ϕ) are the spherical polar coordinates of the scattered particle. Thus for large r, $u(\mathbf{r})$ must satisfy the asymptotic boundary condition

$$u(\mathbf{r}) \; \longrightarrow \; e^{ikz} + f(k,\theta,\phi)\frac{e^{ikr}}{r}. \tag{14.18}$$

$f(k,\theta,\phi)$ is called the *scattering amplitude* .

Now, the probability current density for the stationary state Eq. (14.8) is given by

$$\boldsymbol{j}(\mathbf{r}) \; = \; -i\frac{\hbar}{2m}\left[u^*(\mathbf{r})\,\boldsymbol{\nabla}u(\mathbf{r}) - (\boldsymbol{\nabla}u^*(\mathbf{r}))\,u(\mathbf{r})\right]. \tag{14.19}$$

For large r, the radial current of the scattered particles in the direction of (θ,ϕ) can be obtained from Eq. (14.19) and Eq. (14.18) and is given by

$$j_r \; = \; \frac{k\hbar}{m}\frac{|f(k,\theta,\phi)|^2}{r^2}. \tag{14.20}$$

The number of particles entering the detector per unit time, $Nd\Omega$ is then

$$Nd\Omega \; = \; \frac{k\hbar}{m}|f(k,\theta,\phi)|^2\,d\Omega, \tag{14.21}$$

and the *differential scattering cross section* is given by

$$\frac{d\sigma}{d\Omega} \; = \; \frac{N}{F} \; = \; \frac{N}{v} \; = \; |f(k,\theta,\phi)|^2. \tag{14.22}$$

14.3   The Method of Partial Waves

When the scattering potential is central, i.e. $V(\mathbf{r}) = V(r)$, there is complete symmetry about the direction of incidence (which is taken as the z-axis) so that the wavefunction and hence the scattering amplitude depend on θ but *not* on ϕ, the azimuthal angle. Then $u(r,\theta)$ and $f(k,\theta)$ can be expanded in terms of Legendre polynomials, which form a complete set in the interval of $\cos\theta$ between $+1$ and -1, i.e.

$$-1 \; \leq \; \cos\theta \; \leq \; +1.$$

Thus

$$u(r,\theta) = \sum_{l=0}^{\infty} R_l(k,r) P_l(\cos\theta), \tag{14.23}$$

$$\text{and } f(k,\theta) = \sum_{l=0}^{\infty} f_l(k) P_l(\cos\theta). \tag{14.24}$$

Each term of Eq. (14.23) is called a *partial* wave, which is a simultaneous eigenfunction of the orbital angular momentum operators $\hat{L}^2$ and $\hat{L}_z$ belonging to eigenvalues $l(l+1)\hbar^2$ and 0 respectively. In spectroscopic notation $l = 0, 1, 2, 3, \cdots$ partial waves are known as $s, p, d, f, \cdots$ waves.

The partial wave amplitudes $f_l(k)$ in Eq. (14.24) are determined by the radial function $R_l(k,r)$ as will be known now.

The radial wavefunction $R_l(k,r)$ satisfies the equation

$$\left[\frac{d^2}{dr^2} + \frac{2}{r}\frac{d}{dr} - \frac{l(l+1)}{r^2} - V_{\text{eff}}(r) + k^2\right] R_l(k,r) = 0, \tag{14.25}$$

where $$V_{\text{eff}}(r) = \frac{2m}{\hbar^2}V(r) \text{ and } E = \frac{k^2\hbar^2}{2m}.$$

For potential less singular than r^{-2} at the origin, the behaviour of $R_l(k,r)$ near $r = 0$ can be determined by the power series expansion

$$R_l(k,r) = r^s \sum_{n=0}^{\infty} a_n r^n. \tag{14.26}$$

Indicial equation gives two solutions with $s = l$ and $s = -(l+1)$. Since the wavefunction $u(r,\theta)$ must be finite everywhere including the origin $r = 0$, the solution with $s = l$ is the correct one to keep. In the other case for $s = -(l+1)$ the wavefunction blows up at $r = 0$. For sufficiently large r, say for $r > d$, the potential $V_{\text{eff}}(r)$ can be neglected and the equation satisfied by $R_l(k,r)$ is given by

$$\left[\frac{d^2}{dr^2} + \frac{2}{r}\frac{d}{dr} - \frac{l(l+1)}{r^2} + k^2\right] R_l(k,r) = 0. \tag{14.27}$$

The linearly independent solutions of Eq. (14.27) are the *spherical Bessel* and *spherical Neumann* functions $j_l(kr)$ and $\eta_l(kr)$. The general solution will be a linear combination of these functions, so that the radial function $R_l(k,r)$ in the region $r > d$(where $V_{\text{eff}}(r)$ can be neglected) is given by

$$R_l(k,r) = B_l(k) j_l(kr) + C_l(k) \eta_l(kr), \text{ for } r > d. \tag{14.28}$$

Using the asymptotic expressions for $j_l(kr)$ and $\eta_l(kr)$ we can write

$$\lim_{r \to \infty} R_l(k,r) \longrightarrow \frac{1}{kr}\left[B_l(k)\sin\left(kr - \frac{l\pi}{2}\right) - C_l(k)\cos\left(kr - \frac{l\pi}{2}\right)\right], \quad (14.29)$$

which can be re-written as

$$\lim_{r \to \infty} R_l(k,r) \longrightarrow A_l(k)\frac{1}{kr}\sin\left(kr - \frac{l\pi}{2} + \delta_l(k)\right) \quad (14.30)$$

$$\text{where } A_l(k) = \sqrt{B_l^2(k) + C_l^2(k)}, \quad (14.31)$$

$$\text{and } \delta_l(k) = -\arctan\left[\frac{C_l(k)}{B_l(k)}\right]. \quad (14.32)$$

The real constants $\delta_l(k)$ are called *phase shifts* and characterize the strength of the scattering in the l-th partial wave by V_{eff} at energy $E = \frac{k^2\hbar^2}{2m}$. This is so, because if V_{eff} is zero, the physical solution of Eq. (14.25), that is the solution which behaves like r^l at the origin. is the function $j_l(kr)$ which has the asymptotic form Eq. (14.30) with $\delta_l(k) = 0$.

Since f is independent of ϕ, the asymptotic form of $u(r,\theta)$ for latge r is

$$\lim_{r \to} u(r,\theta) \longrightarrow e^{ikz} + f(k,\theta)\frac{e^{ikr}}{r}, \quad (14.33)$$

and expanding e^{ikz} in terms of $P_l(\cos\theta)$

$$e^{ikz} = \sum_{l=0}^{\infty}(2l+1)i^l j_l(kr)P_l(\cos\theta) \quad (14.34)$$

$$\longrightarrow \sum_{l=0}^{\infty}(2l+1)i^l\frac{1}{kr}\sin\left(kr - \frac{l\pi}{2}\right)P_l(\cos\theta). \quad (14.35)$$

Thus

$$\lim_{r \to \infty} R_l(k,r) \longrightarrow (2l+1)i^l\frac{1}{kr}\sin\left(kr - \frac{l\pi}{2}\right) + \frac{1}{r}f_l(k). \quad (14.36)$$

Equating Eq. (14.36) with Eq. (14.30) we have

$$(2l+1)i^l\frac{1}{kr}\sin\left(kr - \frac{l\pi}{2}\right) + \frac{1}{r}f_l(k)\exp(ikr) =$$

$$A_l(k)\frac{1}{kr}\sin\left(kr - \frac{l\pi}{2} + \delta_l(k)\right). \quad (14.37)$$

Problem 14.2 From Eq. (14.37) show the following

$$A_l(k) = (2l+1) i^l \exp\left[i\delta_l(k)\right] \tag{14.38}$$

$$f_l(k) = (2l+1) \frac{1}{k} e^{i\delta_l(k)} \sin\left[\delta_l(k)\right]. \tag{14.39}$$

Hint: Extress Sine terms as complex exponentials and equate separately coefficints of e^{+ikr} and e^{-ikr} from both sides.

Thus we have

$$f(k,\theta) = \sum_{l=0}^{\infty} f_l(k) P_l(\cos\theta)$$

$$= \frac{1}{2ik} \sum_{l=0}^{\infty} (2l+1) \left\{ \exp\left[2i\delta_l(k) - 1\right] \right\} P_l(\cos\theta), \tag{14.40}$$

which depends only on the phase shifts $\delta_l(k)$ and not on the normalization of the radial function given by $A_l(k)$ in Eq. (14.38).

To understand the significance of the phase shifts we write the asymptotic form of $R_l(k,r)$ given by Eq. (14.25)

$$\lim_{r\to\infty} R_l(k,r) \longrightarrow \frac{1}{2ik} A_l(k) e^{-i\delta_l(k)} \left[\frac{e^{-i\left(kr-\frac{l\pi}{2}\right)}}{r} - S_l(k) \frac{e^{+i\left(kr-\frac{l\pi}{2}\right)}}{r} \right] \tag{14.41}$$

$$\text{where } S_l(k) = \exp\left[2i\delta_l(k)\right] \tag{14.42}$$

The first term on the right hand side in Eq. (14.41) represents an incoming spherical wave and the second term an outgoing spherical wave. $S_l(k)$ is called the *S-matrix element* whose modulus is unity in the present case. This is so, because in the elastic scattering the number of particles entering the scattering region per second must be equal to the number of particles leaving the region per second. As a result the effect of the potential is to produce a phase difference between the ingoing and the outgoing spherical waves. The S-matrix can be generalized to describe non-elastic scattering process as well. Then the conservation of probability is expressed by requiring S-matrix to be unitary.

THE TOTAL CROSS SECTION

This can be calculated integrating the differential cross section Eq. (14.22)

$$\sigma_{\text{tot}} = \int d\sigma = \int |f(k,\theta)|^2 \, d\Omega = 2\pi \int_{-1}^{+1} d(\cos\theta) \, f^*(k,\theta) \, f(k,\theta). \tag{14.43}$$

Using Eq. (14.40) and the orthogonality of the Legendre polynomials:

$$\int_{-1}^{+1} d\left(\cos\theta\right) P_l\left(\cos\theta\right) P_{l'}\left(\cos\theta\right) \;=\; \frac{2}{2l+1}\delta_{l,l'}. \tag{14.44}$$

we get

$$\sigma_{\text{tot}} \;=\; \sum_{l=0}^{\infty} \sigma_l \tag{14.45}$$

$$\text{where } \sigma_l \;=\; \frac{4\pi}{2l+1}\left|f_l\left(k\right)\right|^2$$

$$\;=\; \frac{4\pi}{k^2}\left(2l+1\right)\sin^2\delta_l. \tag{14.46}$$

Problem 14.3 Use Eq. (14.39), Eq. (14.40) Eq. (14.43) and Eq. (14.44) to prove Eq. (14.45) and Eq. (14.46).

14.4   The Optical Theorem

We shall now show that the total scattering cross section is related to the forward angle scattering amplitude $f\left(k,\theta=0\right)$.

From the generating function of Legendre polynomial it follows that $P_l\left(1\right)=1$ for all l; so from Eq. (14.40) we have

$$f\left(k,\theta=0\right) \;=\; \frac{1}{2ik}\sum_{l=0}^{\infty}\left(e^{2i\delta_l}-1\right)$$

Now

$$\frac{2\pi}{ik}\left[f\left(k,\theta=0\right)-f^*\left(k,\theta=0\right)\right] \;=\; \frac{4\pi}{k^2}\sum_{l=0}^{\infty}\left(2l+1\right)\sin^2\delta_l \;=\; \sigma_{\text{tot}}.$$

$$\tag{14.47}$$

$$\therefore \qquad\qquad \sigma_{\text{tot}} \;=\; \frac{4\pi}{k}\Im f\left(k,\theta=0\right). \tag{14.48}$$

where $\Im$ denotes the imaginary part of the argument. This relation Eq. (14.48) is known a *optical theorem* that connects the total cross section to the forward angle scattering amplitude.

To understand the physical significance of Eq. (14.48) we observe that in order for scattering to take place, particles must be removed by an amount proportional to σ from the incident beam, so that its intensity is smaller behind the scattering region ($\theta \approx 0$) than in front of it. This occurs only by interference between the two terms in the asymptotic expression Eq. (14.18). It can be shown by actual calculation of this interference term that optical theorem holds much more generally, namely when $f(k)$ depends on ϕ as well as on θ and σ includes inelastic scattering and absorption as well as elastic scattering.

CONVERGENCE OF PARTIAL WAVE SERIES

The expression of σ_{tot} in Eq. (14.45) or that of $f(k, \theta)$ are useless unless the series in l converges rapidly. We see in Eq. (14.25) that as l increases the centrifugal barrier becomes more important than $V_{\text{eff}}(r)$, so for sufficiently large l, V_{eff} can be neglected and the corresponding phase shift δ_l or the partial wave amplitude f_l is negligible. If $l_{\max}$ is the maximum of l which contributes to the series then $l_{\max}$ increases as the energy increases. We can estimate the number of important partial waves following a semiclassical arguments as follows. If the range of potential is a, beyond which it vanishes then an incident particle having an *impact parameter* 'b' will be deflected or not according to whether $b < a$ or $b > a$.

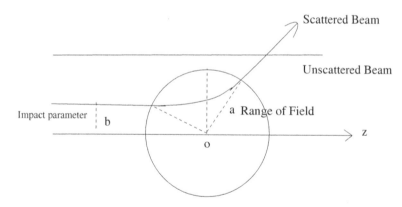

Figure 14.4: Geometrical depiction of Impact Parameter b and range a of scattering potential.

The magnitude of angular momentum L in classical mechanics of a particle of linear momentum p is $L = p \cdot b$, so that if $L > p \cdot a$ then the particle will be undeflected. In the limit of large l, L can be set equal to $l\hbar$, and setting $p = k\hbar$, we see that scattering in the l-th partial wave is expected to be small if

$$b > ka. \tag{14.49}$$

It can also be shown that in the limit of $k \to 0$ only $l = 0$ wave is of importance.

Scattering length may be defined as

$$\alpha \quad = \quad -\lim_{k \to 0} \frac{1}{k} \tan \delta_0, \tag{14.50}$$

and $\lim_{k \to 0} f \quad \longrightarrow \quad -\alpha,$ $\qquad(14.51)$

$$\frac{d\sigma}{d\Omega} \quad = \quad \alpha^2, \qquad \sigma_{\text{tot}} \quad = \quad 4\pi\alpha^2. \tag{14.52}$$

Chapter 15

Scattering Theory II

15.1 The Lippmann Schwinger Equation

In this chapter we deal with formal theory of scattering. We start with the scattering process in a time independent formulation. The Hamiltonian is assumed to be given as usual by

$$\hat{H} = \hat{H}_0 + \hat{V}. \tag{15.1}$$

where $\hat{H}_0$ is the kinetic energy operator

$$\hat{H}_0 = \frac{\hat{p}^2}{2m}. \tag{15.2}$$

In the absence of the scatterer, $\hat{V}$ would be zero and the energy eigenstates would just be the free particle states $|\mathbf{p}\rangle$. The presence of $\hat{V}$ causes eigenstates to be different from free particle states. For elastic scattering there is no change in energy and we are interested in obtaining a solution to the full Hamiltonian Schrödinger equation with the same energy eigenvalue. Thus if $|\phi\rangle$ be the energy eigenket of $\hat{H}_0$

$$\hat{H}_0|\phi\rangle = E|\phi\rangle, \tag{15.3}$$

then the basic Schrödinger equation we wish to solve is

$$\left(\hat{H}_0 + \hat{V}\right)|\psi\rangle = E|\psi\rangle. \tag{15.4}$$

Here, we denote the eigenstate of $\hat{H}_0$ by $|\phi\rangle$ instead of $|\mathbf{p}\rangle$, because we may later be interested in free spherical wave rather than plane wave state. $|\phi\rangle$ may stand for either. Both $\hat{H}_0$ and $\hat{H}_0 + \hat{V}$ exhibit continuous energy spectra and we look for a solution to Eq. (15.4) such that $|\psi\rangle \rightarrow |\phi\rangle$ as $\hat{V} \rightarrow 0$, where $|\phi\rangle$ is the solution of the free particle Schrödinger equation Eq. (15.3) with the same energy. Formally, we may argue that the desired solution is

$$|\psi\rangle = \frac{1}{E - \hat{H}_0}\hat{V}|\psi\rangle + |\phi\rangle, \tag{15.5}$$

if for the time being we ignore the fact that the operator $\frac{1}{E-\hat{H}_0}$ is of singular nature. We note that $\left(E - \hat{H}_0\right)$, applied to Eq. (15.5) immediately yields Eq. (15.4). To obtain the right equation one makes E slightly complex and the correct equation is given by

$$|\psi^\pm\rangle = |\phi\rangle + \frac{1}{E - \hat{H}_0 \pm i\epsilon}\hat{V}|\psi^\pm\rangle. \tag{15.6}$$

This is known as the *Lippmann Schwinger equation*. The physical meaning of the $\pm$ sign will become evident by looking at $\langle\mathbf{x}|\psi^\pm\rangle$ at large distances. Eq. (15.6) in position representation becomes

$$\begin{aligned}
\langle\mathbf{x}|\psi^\pm\rangle &= \langle\mathbf{x}|\phi\rangle + \left\langle\mathbf{x}\left|\frac{1}{E - \hat{H}_0 \pm i\epsilon}\hat{V}\right|\psi^\pm\right\rangle \\
&= \langle\mathbf{x}|\phi\rangle + \int d^3\mathbf{x}'\left\langle\mathbf{x}\left|\frac{1}{E - \hat{H}_0 \pm i\epsilon}\right|\mathbf{x}'\right\rangle\left\langle\mathbf{x}'\left|\hat{V}\right|\psi^\pm\right\rangle,
\end{aligned} \tag{15.7}$$

where we introduced the unit operator $\int d^3\mathbf{x}'|\mathbf{x}'\rangle\langle\mathbf{x}'|$. This is an integral equation for scattering because the unknown ket $|\psi^\pm\rangle$ appears under the integral sign. If $|\mathbf{p}\rangle$ stands for a plane wave state with momentum $\mathbf{p}$, we can write

$$\langle\mathbf{x}|\phi\rangle = \frac{e^{i\mathbf{p}\cdot\mathbf{x}/\hbar}}{(2\pi\hbar)^{\frac{3}{2}}}. \tag{15.8}$$

Unlike the bound states, the plane wave states Eq. (15.8) are not normalizable and are not really a vector in Hilbert space. We have delta function normalization in this case given by

$$\int d^3\mathbf{x}\langle\mathbf{p}'|\mathbf{x}\rangle\langle\mathbf{x}|\mathbf{p}\rangle = \delta^3\left(\mathbf{p} - \mathbf{p}'\right). \tag{15.9}$$

However, in momentum basis the Lippmann-Schwinger equation takes the form

$$\langle\mathbf{p}|\psi^\pm\rangle = \langle\mathbf{p}|\phi\rangle + \frac{1}{E - \frac{p^2}{2m} \pm i\epsilon}\langle\mathbf{p}|\hat{V}|\psi^\pm\rangle. \tag{15.10}$$

We now proceed to evaluate the kernel of the integral equation Eq. (15.7) defined by

$$G_{\pm}(\mathbf{x}, \mathbf{x}') \equiv \frac{\hbar^2}{2m} \left\langle \mathbf{x} \left| \frac{1}{E - \hat{H}_0 \pm i\epsilon} \right| \mathbf{x}' \right\rangle. \tag{15.11}$$

It can be shown that

$$G_{\pm}(\mathbf{x}, \mathbf{x}') = -\frac{1}{4\pi} \frac{e^{\pm ik|\mathbf{x}-\mathbf{x}'|}}{|\mathbf{x} - \mathbf{x}'|}, \tag{15.12}$$

where $E = \frac{k^2 \hbar^2}{2m}$.

To evaluate Eq. (15.11) we proceed as follows:

$$\frac{\hbar^2}{2m} \left\langle \mathbf{x} \left| \frac{1}{E - \hat{H}_0 \pm i\epsilon} \right| \mathbf{x}' \right\rangle = \frac{\hbar^2}{2m} \int d^3p' \int d^3p'' \left\langle \mathbf{x} \mid \mathbf{p}' \right\rangle \times$$
$$\left\langle \mathbf{p}' \left| \frac{1}{E - \frac{p'^2}{2m} \pm i\epsilon} \right| \mathbf{p}'' \right\rangle \left\langle \mathbf{p}'' \mid \mathbf{x}' \right\rangle, \tag{15.13}$$

where $\hat{H}_0$ has acted on $\langle \mathbf{p}'|$.

Now

$$\left\langle \mathbf{p}' \left| \frac{1}{E - \frac{p'^2}{2m} \pm i\epsilon} \right| \mathbf{p}'' \right\rangle = \frac{\delta^3(\mathbf{p}' - \mathbf{p}'')}{E - \frac{p'^2}{2m} \pm i\epsilon}, \tag{15.14}$$

$$\langle \mathbf{x} | \mathbf{p}' \rangle = \frac{e^{i\mathbf{p}' \cdot \mathbf{x}/\hbar}}{(2\pi\hbar)^{3/2}}, \qquad \langle \mathbf{p}'' | \mathbf{x}' \rangle = \frac{e^{-i\mathbf{p}'' \cdot \mathbf{x}'/\hbar}}{(2\pi\hbar)^{3/2}}. \tag{15.15}$$

Then the right hand side of Eq. (15.13) becomes

$$\frac{\hbar^2}{2m} \int \frac{d^3p'}{(2\pi\hbar)^3} \frac{\exp\left[i\frac{\mathbf{p}' \cdot (\mathbf{x}-\mathbf{x}')}{\hbar}\right]}{E - \frac{p'^2}{2m} \pm i\epsilon}.$$

We write $E = \frac{k^2\hbar^2}{2m}$ and set $\mathbf{p}' \equiv \mathbf{q}\hbar$ to obtain

$$\frac{1}{(2\pi)^3} \int_0^\infty q^2 dq \int_0^{2\pi} d\phi \int_{-1}^{+1} d(\cos\theta) \frac{\exp[iq|\mathbf{x} - \mathbf{x}'|\cos\theta]}{k^2 - q^2 \pm i\epsilon} =$$
$$-\frac{1}{8\pi^2} \frac{1}{|\mathbf{x} - \mathbf{x}'|} \int_{-\infty}^{+\infty} dq \frac{q\left[e^{+iq|\mathbf{x}-\mathbf{x}'|} - e^{-iq|\mathbf{x}-\mathbf{x}'|}\right]}{-k^2 + q^2 \mp i\epsilon} = I.$$

$$\tag{15.16}$$

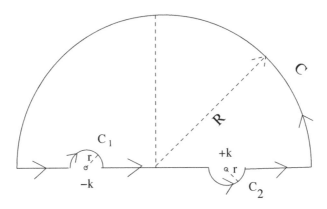

Figure 15.1: The contour in the complex q-plane for evaluating Eq. (15.16).

To evaluate the integral in Eq. (15.16) we note that that the integrand has poles in the complex q-plane at

$$q = -k - i\epsilon, \text{ and } q = k + i\epsilon. \tag{15.17}$$

We choose a contour as in Fig. 15.1. Since the integrals on C_1, C_2 and C can be shown to be zeroes in the limit $R \to \infty$ and $r \to 0$, we have from the method of residues I in Eq. 15.16 becomes

$$I = -\frac{1}{4\pi} \frac{\exp\left[\pm ik\left|\mathbf{x} - \mathbf{x}'\right|\right]}{\left|\mathbf{x} - \mathbf{x}'\right|}. \tag{15.18}$$

Thus the kernel $G_\pm$ in Eq. (15.11) is the same as the Green's function for the Helmholtz equation

$$\left(\boldsymbol{\nabla}^2 + k^2\right) G_\pm\left(\mathbf{x}, \mathbf{x}'\right) = \delta^3\left(\mathbf{x} - \mathbf{x}'\right). \tag{15.19}$$

Finally we get from Eq. (15.7)

$$\langle\mathbf{x}|\psi^\pm\rangle = \langle\mathbf{x}|\phi\rangle - \frac{2m}{\hbar^2}\int d^3\mathbf{x}'\frac{\exp\left[\pm ik\left|\mathbf{x} - \mathbf{x}'\right|\right]}{4\pi\left|\mathbf{x} - \mathbf{x}'\right|}\langle\mathbf{x}'|\hat{V}|\psi^\pm\rangle. \tag{15.20}$$

The wavefunction $\langle\mathbf{x}|\psi^\pm\rangle$ in the presence of the scatterer can thus be written as the sum of the incident wave and a term representing the effect of the scattering through $\hat{V}$.

At large distances the spatial dependence of the second term can be shown to be $\frac{\exp(\pm ikr)}{r}$ provided that the potential is of finite range. This means that the positive solution (negative solution) corresponds to the plane wave plus an outgoing (incoming) spherical wave. We are here interested in the positive solution. If the potential is local which means it is diagonal in x-representation, then

$$\langle \mathbf{x}'|\hat{V}|\mathbf{x}''\rangle = V(\mathbf{x}')\,\delta^3(\mathbf{x}'-\mathbf{x}'') \tag{15.21}$$

$$\text{and } \langle \mathbf{x}'|\hat{V}|\psi^\pm\rangle = \int d^3\mathbf{x}'' \langle \mathbf{x}'|\hat{V}|\mathbf{x}''\rangle\langle \mathbf{x}''|\psi^\pm\rangle = V(\mathbf{x}')\langle \mathbf{x}'|\psi^\pm\rangle. \tag{15.22}$$

Eq. (15.20) now becomes

$$\langle \mathbf{x}|\psi^\pm\rangle = \langle \mathbf{x}|\phi\rangle - \frac{2m}{\hbar^2}\int d^3\mathbf{x}'\frac{\exp(\pm ik|\mathbf{x}-\mathbf{x}'|)}{|\mathbf{x}-\mathbf{x}'|}V(\mathbf{x}')\langle \mathbf{x}'|\psi^\pm\rangle. \tag{15.23}$$

We try to understand the physical content of the equation. The vector $\mathbf{x}$ is naturally directed towards observation point where the wavefunction is evaluated. The source point $\mathbf{x}'$ is confined in the bounded region for a finite range potential, where potential is non-zero. The detector is placed always very far away from the scatterer at r which is greatly larger than the range of the potential.

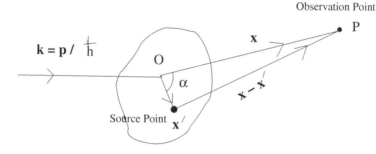

Figure 15.2: The geometry of scattering. We have $r = |\mathbf{x}|$, $r' = |\mathbf{x}'|$ and $r \gg r'$. α is the angle between $\mathbf{x}$ and $\mathbf{x}'$.

Thus for $r \gg r'$

$$\begin{aligned}
|\mathbf{x}-\mathbf{x}'| &= \sqrt{r^2 - 2rr'\cos\alpha + r'^2} \\
&= r\left(1 - 2\frac{r'}{r}\cos\alpha - \frac{r'^2}{r^2}\right)^{\frac{1}{2}} \\
&= r - r'\cos\alpha \\
&= r - \hat{\mathbf{x}}\cdot\mathbf{x}',
\end{aligned} \tag{15.24}$$

where $\hat{\mathbf{x}} \equiv$ unit vector along $\mathbf{x} = \dfrac{\mathbf{x}}{|\mathbf{x}|}$. $\tag{15.25}$

Also we define $\mathbf{k}' \equiv k\hat{\mathbf{x}}$ so that for large r

$$e^{\pm ik|\mathbf{x}-\mathbf{x}'|} = e^{\pm ikr}\cdot e^{\mp i\mathbf{k}'\cdot\mathbf{x}'}. \tag{15.26}$$

Replacing $\frac{1}{|\mathbf{x}-\mathbf{x}'|}$ by $\frac{1}{r}$ we use $|\mathbf{k}\rangle$ instead of $|\mathbf{p}\rangle$ where $\mathbf{k} = \frac{\mathbf{p}}{\hbar}$.

Because $|\mathbf{k}\rangle$ is normalized as $\langle \mathbf{k}|\mathbf{k}'\rangle = \delta^3(\mathbf{k} - \mathbf{k}')$ we have

$$\langle \mathbf{x}|\mathbf{k}\rangle = \frac{e^{i\mathbf{k}\cdot\mathbf{x}}}{(2\pi)^{\frac{3}{2}}}.$$

Finally at $r \to \infty$

$$\langle \mathbf{x}|\psi^+\rangle \longrightarrow \langle \mathbf{x}|\mathbf{k}\rangle - \frac{1}{2\pi}\frac{2m}{\hbar^2}\frac{e^{ikr}}{r}\int d^3(\mathbf{x}')\, e^{-i\mathbf{k}'\cdot\mathbf{x}'}V(\mathbf{x}')\langle \mathbf{x}'|\psi^+\rangle$$

$$= \frac{1}{(2\pi)^{\frac{3}{2}}}\left[e^{i\mathbf{k}\cdot\mathbf{x}} + \frac{e^{ikr}}{r}f(\mathbf{k},\mathbf{k}')\right], \tag{15.27}$$

where the amplitude of the outgoing spherical wave is

$$f(\mathbf{k},\mathbf{k}') = -\frac{1}{4\pi}\frac{2m}{\hbar^2}(2\pi)^3\int d^3(\mathbf{x}')\frac{e^{-i\mathbf{k}'\cdot\mathbf{x}'}}{(2\pi)^{\frac{3}{2}}}V(\mathbf{x}')\langle \mathbf{x}'|\psi^+\rangle$$

$$= -\frac{1}{4\pi}(2\pi)^3\frac{2m}{\hbar^2}\langle \mathbf{k}'|V|\psi^+\rangle. \tag{15.28}$$

Similarly it can readily be shown that $\langle \mathbf{x}|\psi^-\rangle$ corresponds to the original plane wave in the direction of $\mathbf{k}$ plus an incoming spherical wave with

spatial dependence $\dfrac{e^{-ikr}}{r}$ and amplitude $-\dfrac{1}{4\pi}(2\pi)^3\dfrac{2m}{\hbar^2}\langle \mathbf{k}'|V|\psi^-\rangle$.

The differential scattering cross section $\frac{d\sigma}{d\Omega}$ is defined as follows

$$\frac{d\sigma}{d\Omega}d\Omega = \frac{\text{No. of particles scattered in } d\Omega \text{ per unit time}}{\text{No. of incident particle crossing unit area per unit time}} \tag{15.29}$$

$$= \frac{r^2|\mathbf{j}_{sc}|d\Omega}{|\mathbf{j}_{inc}|}$$

$$= |f(\mathbf{k},\mathbf{k}')|^2 d\Omega \tag{15.30}$$

where $\mathbf{j}_{sc}$ and $\mathbf{j}_{inc}$ can be calculated from the definition of probability flux

$$\mathbf{j} = \frac{\hbar}{2im}[\psi^*\boldsymbol{\nabla}\psi - (\boldsymbol{\nabla}\psi^*)\psi] \tag{15.31}$$

using the scattered and incident wavefunctions respectively from Eq. (15.27).

15.2 The Born Approximation

The expression for the scattering amplitude in Eq. (15.28) is not useful as it contains the unknown ket $|\psi^+\rangle$. If the scattering is not very strong, we can replace $\langle \mathbf{x}'|\psi^+\rangle$ appearing under the integral sign by $\langle \mathbf{x}'|\phi\rangle$ as follows

$$\langle \mathbf{x}'|\psi^+\rangle \longrightarrow \langle \mathbf{x}'|\phi\rangle = \frac{e^{i\mathbf{k}\cdot\mathbf{X}'}}{(2\pi)^{\frac{3}{2}}}. \tag{15.32}$$

The *first order Born approximation* is given by

$$f^{(1)}(\mathbf{k},\mathbf{k}') = -\frac{1}{4\pi}\frac{2m}{\hbar^2}\int d^3\mathbf{x}' e^{i(\mathbf{k}-\mathbf{k}')\cdot\mathbf{x}'} V(\mathbf{x}'). \tag{15.33}$$

The integral is the three dimensional Fourier transform of the potential V with respect to $\mathbf{q}$ and $q = |\mathbf{k} - \mathbf{k}'|$. For elastic scattering $|\mathbf{k}'| = |\mathbf{k}|$ and $q \equiv |\mathbf{k} - \mathbf{k}'| = 2k\sin\theta/2$. The angular integration can be carried out explicitly and

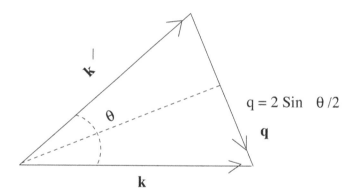

Figure 15.3: Transfer of momentum in a scattering through angle θ. $\mathbf{q} = \mathbf{k} - \mathbf{k}'$.

$$
\begin{aligned}
f^{(1)}(\theta) &= -\frac{1}{2}\frac{2m}{\hbar^2}\frac{1}{iq}\int_0^\infty \frac{V(r)}{r}\left[e^{iqr} - e^{-iqr}\right] r^2 dr \\
&= -\frac{2m}{\hbar^2}\frac{1}{q}\int_0^\infty rV(r)\sin qr\; dr.
\end{aligned} \tag{15.34}
$$

15.3 The Higher Order Born Approximation

We use the *transition operator* T defined by the following

$$\hat{V}|\psi^+\rangle = \hat{T}|\phi\rangle. \tag{15.35}$$

Multiplying the Lippmann-Schwinger's Eq. (15.6) by $\hat{V}$,

$$\hat{T}|\phi\rangle \;=\; \hat{V}|\psi^{+}\rangle \;=\; \hat{V}|\phi\rangle + \hat{V}\frac{1}{E-\hat{H}_0+i\epsilon}\hat{T}|\phi\rangle. \tag{15.36}$$

Since $|\phi\rangle$-s are momentum eigenstates which form complete set, Eq. (15.36) can be written in operator form

$$\hat{T} \;=\; \hat{V} + \hat{V}\frac{1}{E-\hat{H}_0+i\epsilon}\hat{T}. \tag{15.37}$$

We now write the scattering amplitude $f(\mathbf{k},\mathbf{k}')$ from Eq. (15.28) and Eq. (15.35) with $|\phi\rangle$ as momentum eigenkets

$$f(\mathbf{k},\mathbf{k}') \;=\; -\frac{1}{4\pi}\frac{2m}{\hbar^2}(2m)^3\,\langle\mathbf{k}'|\hat{T}|\mathbf{k}\rangle. \tag{15.38}$$

Now the iterative solution for $\hat{T}$ is

$$\begin{aligned}
\hat{T} \;=\;& \hat{V} + \hat{V}\frac{1}{E-\hat{H}_0+i\epsilon}\hat{V} + \hat{V}\frac{1}{E-\hat{H}_0+i\epsilon}\hat{V}\frac{1}{E-\hat{H}_0+i\epsilon}\hat{V} + \cdots \\
\;=\;& \left(\sum_{n=1}^{\infty}\left[\hat{V}\frac{1}{E-\hat{H}_0+i\epsilon}\right]^{n-1}\right)\hat{V}
\end{aligned} \tag{15.39}$$

and correspondingly the expansion of $f(\mathbf{k},\mathbf{k}')$ is

$$\begin{aligned}
f(\mathbf{k},\mathbf{k}') \;=\;& \sum_{n=1}^{\infty} f^{(n)}(\mathbf{k},\mathbf{k}') \\
\text{with } f^{(n)}(\mathbf{k},\mathbf{k}') \;=\;& -\frac{1}{4\pi}\frac{2m}{\hbar^2}(2\pi)^3\left\langle\mathbf{k}'\left|\hat{V}\left[\frac{1}{E-\hat{H}_0+i\epsilon}\hat{V}\right]^{n-1}\right|\mathbf{k}\right\rangle.
\end{aligned} \tag{15.40}$$

where n is the number of $\hat{V}$ operators appearing in the expression and also denotes the order of Born approximation. Thus

$$\begin{aligned}
f^{(1)}(\mathbf{k},\mathbf{k}') \;=\;& -\frac{1}{4\pi}\frac{2m}{\hbar^2}(2\pi)^3\left\langle\mathbf{k}'\left|\hat{V}\right|\mathbf{k}\right\rangle \\
f^{(2)}(\mathbf{k},\mathbf{k}') \;=\;& -\frac{1}{4\pi}\frac{2m}{\hbar^2}(2\pi)^3\left\langle\mathbf{k}'\left|\hat{V}\frac{1}{E-\hat{H}_0+i\epsilon}\hat{V}\right|\mathbf{k}\right\rangle \\
\cdots \;=\;& \cdots
\end{aligned} \tag{15.41}$$

Introducing the identity operator $\int d^3 |\mathbf{x}'\rangle\langle\mathbf{x}'|$ at relevant points and using the locality of $\hat{V}$ we can write

$$
\begin{aligned}
f^{(2)}(\mathbf{k}, \mathbf{k}') &= -\frac{1}{4\pi}\frac{2m}{\hbar^2}(2\pi)^3 \int d^3\mathbf{x}' \int d^3\mathbf{x}'' \langle\mathbf{k}'|\mathbf{x}'\rangle V(\mathbf{x}') \times \\
&\qquad \left\langle \mathbf{x}'\left|\frac{1}{E-\hat{H}_0+i\epsilon}\right|\mathbf{x}''\right\rangle V(\mathbf{x}'') \langle\mathbf{x}''|\mathbf{k}\rangle \\
&= -\frac{1}{4\pi}\frac{2m}{\hbar^2} \int d^3\mathbf{x}' \int d^3\mathbf{x}'' e^{-i\mathbf{k}'\cdot\mathbf{x}'} V(\mathbf{x}') \left[\frac{2m}{\hbar^2}G_+(\mathbf{x}',\mathbf{x}'')\right] \times \\
&\qquad V(\mathbf{x}'') e^{+i\mathbf{k}\cdot\mathbf{x}''}
\end{aligned} \tag{15.42}
$$

Second Born Approximation

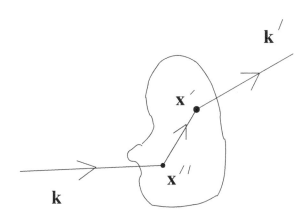

Figure 15.4: Schematic description of 2nd order Born scattering process.

Physically Eq. (15.42) can be interpreted using Fig. 15.4 as follows. The incident wave interacts at $\mathbf{x}''$ since $V(\mathbf{x}'')$ is appearing, and then propagates to $\mathbf{x}'$ via the Green's function $G_+(\mathbf{x}',\mathbf{x}'')$ and subsequently a second interaction occurs at $\mathbf{x}'$, because of $V(\mathbf{x}')$. Finally the wave is scattered in the direction of $\mathbf{k}'$. Thus $f^{(2)}$ is a two step process, likewise $f^{(3)}$ and so on.

Chapter 16

Relativistic Wave Equations

16.1   Introduction

A relativistic wave equation describes the motion of a particle that has speed approaching that of light. A characteristic feature of relativistic wave equation is that the spin of the particle is built into the theory from the start and cannot be added afterwards as Pauli added electron spin to Schrödinger non-relativistic equation. In the next section is described a spin zero particle known as Klein Gordon equation. The subsequent section will deal with Dirac equation describing spin 1/2 particles. We shall continue to use three dimensional vector notations rather than the the four dimensional notation of special relativity theory. However, the invariance of an equation under Lorentz transformation can usually be inferred from its symmetry between the space coordinate and time.

16.2   The Klein Gordon Equation

This equation follows from the relativistic energy expression for a free particle of mass m.

$$E^2 = p^2c^2 + m^2c^4. \tag{16.1}$$

Using the operators

$$E = i\hbar\frac{\partial}{\partial t} \quad \text{and} \quad \boldsymbol{p} = -i\hbar\boldsymbol{\nabla}, \tag{16.2}$$

Eq. (16.1) becomes

$$-\frac{\partial^2\psi}{\partial t^2} = -\hbar^2c^2\boldsymbol{\nabla}^2\psi + m^2c^4\psi. \tag{16.3}$$

Eq. (16.3) has plane wave solutions of the form

$$\psi\left(\mathbf{r},t\right) \;=\; A\exp\left[i\left(\mathbf{k}\cdot\mathbf{r}-\omega t\right)\right], \tag{16.4}$$

which are eigenfunctions of the operators E and $\boldsymbol{p}$ in Eq. (16.2) with eigenvalues $\hbar\omega$ and $\hbar\boldsymbol{p}$ respectively.

Now Eq. (16.4) satisfies Eq. (16.3) if

$$\hbar\omega \;=\; \pm\sqrt{\hbar^2 c^2 \boldsymbol{k}^2 + m^2 c^4}. \tag{16.5}$$

The $\pm$ in Eq. (16.5) is due to the ambiguity in the sign of the energy resulting from the classical expressiom Eq. (16.1)

The continuity equation

$$\frac{\partial P\left(\boldsymbol{r},t\right)}{\partial t} + \boldsymbol{\nabla}\cdot\boldsymbol{S}\left(\boldsymbol{r},t\right) \;=\; 0 \tag{16.6}$$

is invariant under Lorentz transformation. Eq. (16.6) results if we define

$$P\left(\boldsymbol{r},t\right) \;=\; \frac{i\hbar}{2mc^2}\left[\psi^*\frac{\partial\psi}{\partial t} - \psi\frac{\partial\psi^*}{\partial t}\right]$$

$$\text{and } \boldsymbol{S}\left(\boldsymbol{r},t\right) \;=\; \frac{\hbar}{2im}\left[\psi^*\boldsymbol{\nabla}\psi - \psi\boldsymbol{\nabla}\psi^*\right]. \tag{16.7}$$

However, $P\left(\boldsymbol{r},t\right)$ defined in Eq. (16.7) is not necessarily positive definite and hence cannot be interpreted as position probability density. But it, when multiplied by the charge e, can be interpreted as charge density, since charge density can have either sign so long as it is real.

The Klein Gordon equation suffers from two defects: lack of existence of a positive definite probability density, and occurrence of negative energy states. For a free particle, whose energy is constant, the latter difficulty may be avoided by choosing only positive energy for the particle and simply ignoring the negative energy states. But an interacting particle may exchange energy with its environment, and there would then be nothing to stop it cascading down to negative energy states emitting an infinite amount of energy in the process. This, of course, is not observed, and so poses a problem for the single particle Klein Gordon equation. For this reason Klein Gordon equation was discarded and Dirac looked for an equation to replace it.

16.3   The Dirac Relativistic Equation

Dirac started with the wave equation

$$i\hbar\frac{\partial\psi\left(\mathbf{r},t\right)}{\partial t} \;=\; \hat{H}\psi\left(\mathbf{r},t\right) \;=\; E\psi\left(\mathbf{r},t\right). \tag{16.8}$$

The classical relativistic Hamiltonian for a free particle should be the positive square-root of the right hand side of Eq. (16.1). Dirac modified the classical relativistic expression for energy and hence for the Hamiltonian in such a way as to make it linear in space derivative (i.e. linear in momentum). Thus the free particle Dirac Hamiltonian is

$$\hat{H} \;=\; c\boldsymbol{\alpha}\cdot\mathbf{p}+\beta mc^2. \tag{16.9}$$

One can justify the linearity of $\mathbf{p}$ and mc^2 in the expression for $\hat{H}$ as follows. In the relativistic limit, m^2c^4 term in Eq. (16.1) can be neglected and $E \sim pc$. In the non-relativistic limit the momentum term can be neglected and $E \sim mc^2$. Thus the Hamiltonian can be a linear combination of pc and mc^2 in regions away from both the limits.

From Eq. (16.9) and Eq. (16.8) one gets the wave equation

$$\left(E - c\boldsymbol{\alpha}\cdot\mathbf{p}-\beta mc^2\right)\psi\left(\mathbf{r},t\right) \;=\; 0, \tag{16.10}$$

i.e. $\left(i\hbar\dfrac{\partial}{\partial t} - c\boldsymbol{\alpha}\cdot\mathbf{p}-\beta mc^2\right)\psi\left(\mathbf{r},t\right) \;=\; 0,$ \hfill (16.11)

We note that the wave equation is justly linear in space and time derivatives.

Since we are considering the motion of a free particle, all points in space time must be equivalent, i.e. the Hamiltonian should not contain terms that depends on the space coordinate or time, as such terms would give rise to force. Also, the space and time derivatives are to appear only in $\mathbf{p}$ and E. We thus conclude that $\boldsymbol{\alpha}$ and β are independent $\mathbf{r}, t, \mathbf{p}$ and E and hence commute with them. However this does not neccessarily mean that $\boldsymbol{\alpha}$ and β are numbers, since they need not commute with each other.

We now require that any solution ψ of Eq. (16.11) shall also be solution of the Klein Gordon relativistic Eq. (16.3), We therefore multiply Eq. (16.10) on the left by

$$\left(E + c\boldsymbol{\alpha}\cdot\mathbf{p}+\beta mc^2\right)$$

and get

$$\begin{aligned}
\bigl\{ E^2 &- c^2\left[\alpha_x^2 p_x^2 + \alpha_y^2 p_y^2 + \alpha_z^2 p_z^2\right] - m^2 c^4\beta^2 \\
&- c^2\left[(\alpha_x\alpha_y + \alpha_y\alpha_x)\,p_x p_y + (\alpha_y\alpha_z + \alpha_z\alpha_y)\,p_y p_z + (\alpha_z\alpha_x + \alpha_x\alpha_z)\,p_z p_x\right] \\
&- mc^3\left[(\alpha_x\beta + \beta\alpha_x)\,p_x + (\alpha_y\beta + \beta\alpha_y)\,p_y + (\alpha_z\beta + \beta\alpha_z)\,p_z\right]\bigr\}\,\psi \;=\; 0,
\end{aligned} \tag{16.12}$$

where E and $\mathbf{p}$ are differential operators given by Eq. (16.2). Eq. (16.11) agrees with Eq. (16.3) if $\boldsymbol{\alpha}, \beta$ satisfy the relations

$$
\begin{aligned}
\alpha_x^2 = \alpha_y^2 = \alpha_z^2 = \beta^2 &= \hat{I}, \\
\alpha_x\alpha_y + \alpha_y\alpha_x = \alpha_y\alpha_z + \alpha_z\alpha_y = \alpha_z\alpha_x + \alpha_x\alpha_z &= 0, \\
\alpha_x\beta + \beta\alpha_x = \alpha_y\beta + \beta\alpha_y = \alpha_z\beta + \beta\alpha_z &= 0.
\end{aligned}
$$

$$(16.13)$$

MATRICES FOR $\boldsymbol{\alpha}$ AND β

It can be shown that they are even dimensional and there are no set of four mutualy anti-commuting 2-dimensional operators. So the minimum dimension is 4. Since the squares of all the four matrices are unity, their eigenvalues are 1 and -1. Thus choosing β arbitrarily diagonal in a particular representation we write

$$
\beta = \begin{pmatrix} \hat{I} & 0 \\ 0 & -\hat{I} \end{pmatrix} = \begin{pmatrix} 1 & 0 & 0 & 0 \\ 0 & 1 & 0 & 0 \\ 0 & 0 & -1 & 0 \\ 0 & 0 & 0 & -1 \end{pmatrix}.
$$

$$(16.14)$$

It can then be shown using the properties of $\boldsymbol{\alpha}$ and β given Eq. (16.13) that

$$
\alpha = \begin{pmatrix} 0 & \sigma \\ \sigma & 0 \end{pmatrix}
$$

$$(16.15)$$

where $\boldsymbol{\sigma}$-s are the 2×2 Pauli spin matrices:

$$
\sigma_x = \begin{pmatrix} 0 & 1 \\ 1 & 0 \end{pmatrix}, \; \sigma_y = \begin{pmatrix} 0 & -i \\ i & 0 \end{pmatrix}, \; \sigma_z = \begin{pmatrix} 1 & 0 \\ 0 & -1 \end{pmatrix}.
$$

Thus

$$
\alpha_x = \begin{pmatrix} 0 & \sigma_x \\ \sigma_x & 0 \end{pmatrix} = \begin{pmatrix} 0 & 0 & 0 & 1 \\ 0 & 0 & 1 & 0 \\ 0 & 1 & 0 & 0 \\ 1 & 0 & 0 & 0 \end{pmatrix},
$$

$$(16.16)$$

$$
\alpha_y = \begin{pmatrix} 0 & \sigma_y \\ \sigma_y & 0 \end{pmatrix} = \begin{pmatrix} 0 & 0 & 0 & -i \\ 0 & 0 & i & 0 \\ 0 & -i & 0 & 0 \\ i & 0 & 0 & 0 \end{pmatrix},
$$

$$(16.17)$$

$$\alpha_z = \begin{pmatrix} 0 & \sigma_z \\ \sigma_z & 0 \end{pmatrix} = \begin{pmatrix} 0 & 0 & 1 & 0 \\ 0 & 0 & 0 & -1 \\ 1 & 0 & 0 & 0 \\ 0 & -1 & 0 & 0 \end{pmatrix}. \tag{16.18}$$

$$\tag{16.19}$$

Problem 16.1 Verify that the four matrices in Eq. (16.14) to Eq. (16.17) satisfy Eq. (16.13).

FREE PARTICLE SOLUTIONS

We now consider the wave equation Eq. (16.11) where $\boldsymbol{\alpha}$ and β are 4×4 matrices which operate on ψ. Thus the wave function ψ itself has to be a column matrix with four rows, namely

$$\psi(\mathbf{r}, t) = \begin{pmatrix} \psi_1(\mathbf{r}, t) \\ \psi_2(\mathbf{r}, t) \\ \psi_3(\mathbf{r}, t) \\ \psi_4(\mathbf{r}, t) \end{pmatrix}. \tag{16.20}$$

The free particle wave function Eq. (16.11) is then equivalent to four simultaneous first order partial differential equations that are linear and homogeneous in the four ψ-s.

Plane wave solutions of the form

$$\psi_j(\mathbf{r}, t) = u_j \exp\left[i\left(\mathbf{k} \cdot \mathbf{r} - \omega t\right)\right], \quad j = 1, 2, 3, 4. \tag{16.21}$$

can be obtained where u_j-s are numbers. These are eigenfunctions of energy and momentum with eigenvalues $\hbar\omega$ and $\mathbf{k}\hbar$ respectively. Substitution of Eq. (16.20) and the matrices for $\boldsymbol{\alpha}$, β in Eq. (16.11) give the following equations for the u_j-s, where $E = \hbar\omega$ and $\mathbf{p} = \mathbf{k}\hbar$ are now numbers:

$$\begin{aligned}
\left(E - mc^2\right) u_1 + 0 \cdot u_2 - cp_z u_3 - c\left(p_x - ip_y\right) u_4 &= 0, \\
0 \cdot u_1 + \left(E - mc^2\right) u_2 - c\left(p_x + ip_y\right) u_3 + cp_z u_4 &= 0, \\
-cp_z u_1 - c\left(p_x - ip_y\right) u_2 + \left(E - mc^2\right) u_3 + 0 \cdot u_4 &= 0, \\
-c\left(p_x + ip_y\right) u_1 + cp_z u_2 + 0 \cdot u_3 + \left(E - mc^2\right) u_4 &= 0,
\end{aligned} \tag{16.22}$$

These algebraic equations are homogeneous in the u_j-s, and hence have non-zero solutions only if the secular determinant of the coefficients is zero.

Problem 16.2   *(i)*. Show that the secular determinant is $[E^2 - m^2c^4 - p^2c^2]$ where E and $\mathbf{p}$ are numbers.

(ii). Choose $E_+ = +\sqrt{p^2c^2 + m^2c^4}$ for any momementum $\mathbf{p}$ and show that the two linearly independent solutions are given by

$$I. \quad u_1 = 1, \; u_2 = 0, \; u_3 = \frac{cp_z}{E_+ + mc^2}, \; u_4 = \frac{c\,(p_x + ip_y)}{E_+ + mc^2},$$

$$(16.23)$$

$$II. \quad u_1 = 0, \; u_2 = 1, \; u_3 = \frac{c\,(p_x - ip_y)}{E_+ + mc^2}, \; u_4 = \frac{-cp_z}{E_+ + mc^2},$$

$$(16.24)$$

Similar solutions can be obtained for negative energy $E_- = -\sqrt{c^2p^2 + m^2c^4}$. These are given by

$$III. \quad u_1 = \frac{cp_z}{E_- - mc^2}, \; u_2 = \frac{c\,(p_x + ip_y)}{E_- - mc^2}, \; u_3 = 1, \; u_4 = 0,$$

$$(16.25)$$

$$IV. \quad u_1 = \frac{c\,(p_x - ip_y)}{E_- - mc^2}, \; u_2 = \frac{-cp_z}{E_- - mc^2}, \; u_3 = 0, \; u_4 = 1$$

$$(16.26)$$

We have $\psi^\dagger\psi = A_\pm^2$ with

$$A_\pm = \sqrt{1 + \frac{c^2p^2}{(E_\pm \pm mc^2)^2}}$$

(the upper set of signs corresponds to cases I and II and the lower set of signs corresponds to the cases III and IV) where $\psi^\dagger$ is the Hermitian adjoint of ψ, and is a row matrix with four columns. Thus the four solutions can be normalized by dividing them by the corresponding A.

We now define three new spin matrices $\sigma'_x, \sigma'_y, \sigma'_z$ as the 4×4

$$\sigma' = \begin{pmatrix} \sigma & 0 \\ 0 & \sigma \end{pmatrix}. \qquad (16.27)$$

We shall very soon see that $\frac{1}{2}\hbar\sigma'$ can be interpreted as the operator that represents spin angular momentum.

CHARGE AND CURRENT DENSITIES

We can obtain a conservation equation by multiplying Eq. (16.11) on the left by $\psi^\dagger$, the Hermitian adjoint equation

$$-i\hbar\frac{\partial \psi^\dagger}{\partial t} - i\hbar c\left(\boldsymbol{\nabla}\psi^\dagger\right)\cdot\boldsymbol{\alpha} - \psi^\dagger\beta mc^2 = 0$$

on the right by ψ, and taking the difference of the two results. We then obtain

$$\frac{\partial}{\partial t}\left(\psi^\dagger\psi\right) + c\boldsymbol{\nabla}\cdot\left(\psi^\dagger\alpha\psi\right) = 0 \tag{16.28}$$

$$\text{or} \qquad \frac{\partial P}{\partial t} + \boldsymbol{\nabla}\cdot\boldsymbol{S}\left(\mathbf{r},t\right) = 0, \tag{16.29}$$

which is the continuity equation Eq. (16.6) with

$$P = \psi^\dagger\psi, \quad \mathbf{S} = c\left(\psi^\dagger\boldsymbol{\alpha}\psi\right). \tag{16.30}$$

Since P is never negative, it can be interpreted as position probability density.

The second of Eq. (16.30) indicate that $c\boldsymbol{\alpha}$ can be interpreted as a particle velocity. To verify this we calculate the time derivative of position vector $\mathbf{r}$ in the Heisenberg picture, using Eq. (16.10)

$$\frac{dx}{dt} = \frac{1}{i\hbar}\left(x\hat{H} - \hat{H}x\right) = \alpha_x. \tag{16.31}$$

SPIN ANGULAR MOMENTUM IN CENTRAL FIELD

We consider the motion of an electron in a central field. The Hamiltonian is

$$\hat{H} = c\boldsymbol{\alpha}\cdot\mathbf{p} + \beta mc^2 + V\left(r\right), \tag{16.32}$$

$$\text{and } i\hbar\frac{\partial\psi}{\partial t} = \hat{H}\psi.$$

We might expect that the orbital angular momentum $\mathbf{L} = \mathbf{r}\times\mathbf{p}$ would be a constant of motion. However it turns out that the total angular momentum rather than the orbital angular is a constant of motion.

Problem 16.3 *(i)*. From Heisenberg equation of motion calculate $\left[L_x, \hat{H}\right]$ to show that

$$i\hbar\frac{dL_z}{dt} = \left[L_x, \hat{H}\right] = -i\hbar c\left(\alpha_z p_y - \alpha_y p_z\right). \tag{16.33}$$

(ii). Show that

$$[\sigma'_x, \alpha_y] = 2i\alpha_z$$
$$\text{and } [\sigma'_x, \alpha_z] = -2i\alpha_y,$$

so that

$$i\hbar\frac{d\sigma'_x}{dt} = \left[\sigma'_x, \hat{H}\right] = 2ic\left(\alpha_z p_y - \alpha_y p_z\right). \tag{16.34}$$

Thus $L_x + \frac{\hbar}{2}\sigma'_x$ and similarly all the three components commute with $\hat{H}$ and are constants of motion. Identifying $\frac{\hbar}{2}\boldsymbol{\sigma}'$ as the *spin angular momentum* we observe that the *total angular momentum* $\mathbf{J} = \mathbf{L} + \mathbf{S}$ is a constant of motion where

$$\mathbf{S} = \frac{\hbar}{2}\boldsymbol{\sigma}' \tag{16.35}$$

is the spin angular momentum.

16.4   Conclusion

We have seen that in Dirac theory of electron the position probability density is positive.

Also Dirac electrons have negative energy states. Dirac's solution to this problem relies on the fact that electrons have spin 1/2 and therefore obey Pauli's exclusion principle. Dirac assumed that the negative energy states are already completely filled up and the exclusion principle prevents any more electron being able to enter the 'sea' of negative energy states. This 'Dirac sea' is the vacuum; so on Dirac's theory, the vacuum is not 'empty', but an infinite sea of negative energy electrons, protons, neutrinos, neutrons and all other spin 1/2 particles!

Now this ingeneous theory makes an important prediction. For suppose there occurs one vacancy in the electron sea - a 'hole' - with energy $-|E|$. An electron with energy E may fill this hole, emitting energy $2E$.

$$\text{e}^- + \text{hole} \longrightarrow \text{energy}, \tag{16.36}$$

so the 'hole' effectively has charge $+e$ and positive energy and is called a positron, the *antiparticle* of electron. This theory thus predicted the existence of antiparticles for all spin 1/2 particles, and in time $e^+, \bar{p}, \bar{n}, \bar{\nu}$ were observed.

Despite the resolution of negative energy states, Dirac equation is no longer a single particle equation, since it describes both particles and antiparticles. It is then only consistent to regard the spinor ψ as a field such that $\psi^\dagger\psi$ gives a measure of the number of particles at a particular space-time point. This field is obviously a quantum field!

Appendix A

Appendix

A.1 Expansion in a Series of Orthonormal Functions

We consider a free particle in a bounded volume of space Ω, whose orthonormal wavefunctions are

$$\psi_{\mathbf{k}}(\mathbf{r}) = \frac{1}{\sqrt{\Omega}} \exp\left[i\left(\mathbf{k} \cdot \mathbf{r}\right)\right]. \tag{A.1}$$

Suppose we wish to approximate a function $f(\mathbf{r})$ by a linear combination of $\psi_{\mathbf{k}}(\mathbf{r})$ with $|\mathbf{k}| < k_0$, *i.e* we wish to find complex numbers $a_{\mathbf{k}}$ such that in some sense

$$f(\mathbf{r}) \approx \sum_{\mathbf{k}} a_{\mathbf{k}} \psi_{\mathbf{k}}(\mathbf{r}) \quad \text{for} \quad \mathbf{r} \quad \text{within} \quad \Omega$$

and where $|\mathbf{k}| < k_0$. We consider t as fixed and omit it for the present from argument of $\psi_{\mathbf{k}}$.

Define

$$\Delta \equiv \int_{\Omega} \left| f - \sum_{|\mathbf{k}|<k_0} a_{\mathbf{k}} \psi_{\mathbf{k}} \right|^2 d^3\mathbf{r}, \tag{A.2}$$

where $\sqrt{\frac{\Delta}{\Omega}}$ is the *root mean square error*.

Δ is obviously a good measure of the precision to which f is approximated by $\sum_{\mathbf{k}} a_{\mathbf{k}} \psi_{\mathbf{k}}$.

We minimize Δ with respect to $a_{\mathbf{k}}$ to get "least" values of $a_{\mathbf{k}}$.

$$
\begin{aligned}
\Delta &= \int_{\Omega} \left(f - \sum_{|\mathbf{k}|<k_0} a_{\mathbf{k}} \psi_{\mathbf{k}} \right)^* \left(f - \sum_{|\mathbf{k}'|<k_0} a_{\mathbf{k}'} \psi_{\mathbf{k}'} \right) d^3\mathbf{r} \\
&= \int_{\Omega} |f|^2 d^3\mathbf{r} - \sum_{|\mathbf{k}|<k_0} a_{\mathbf{k}}^* \int_{\Omega} \psi_{\mathbf{k}}^* f d^3\mathbf{r} - \sum_{|\mathbf{k}'|<k_0} a_{\mathbf{k}'} \int_{\Omega} f^* \psi_{\mathbf{k}'} d^3\mathbf{r} \\
&\qquad + \sum_{|\mathbf{k}|,|\mathbf{k}'|<k_0} a_{\mathbf{k}}^* a_{\mathbf{k}}' \int_{\Omega} \psi_{\mathbf{k}}^* \psi_{\mathbf{k}'} d^3\mathbf{r} \\
&= \int_{\Omega} |f|^2 d^3\mathbf{r} - \sum_{|\mathbf{k}<k_0} \left[a_{\mathbf{k}}^* \int_{\Omega} \psi_{\mathbf{k}}^* f d^3\mathbf{r} + a_{\mathbf{k}} \int_{\Omega} f^* \psi_{\mathbf{k}} d^3\mathbf{r} \right] \\
&\qquad + \sum_{|\mathbf{k}|,|\mathbf{k}'|<k_0} a_{\mathbf{k}}^* a_{\mathbf{k}'} \delta_{\mathbf{k},\mathbf{k}'} \\
&= \int_{\Omega} |f(\mathbf{r})|^2 d^3\mathbf{r} - \sum_{|\mathbf{k}|<k_0} \left| \int_{\Omega} f(\mathbf{r}) \psi_{\mathbf{k}}^*(\mathbf{r}) d^3\mathbf{r} \right|^2 \\
&\qquad + \sum_{|\mathbf{k}|<k_0} \left| a_{\mathbf{k}} - \int_{\Omega} \psi_{\mathbf{k}}^*(\mathbf{r}) f(\mathbf{r}) d^3\mathbf{r} \right|^2
\end{aligned}
\tag{A.3}
$$

Δ will be minimum for

$$
a_{\mathbf{k}} = \int_{\Omega} \psi_{\mathbf{k}}^*(\mathbf{r})) f(\mathbf{r}) d^3\mathbf{r}, \qquad \text{for all} \quad |\mathbf{k}| < k_0.
\tag{A.4}
$$

and its minumum value is

$$
\Delta_{\min} = \int_{\Omega} |f(\mathbf{r})|^2 d^3\mathbf{r} - \sum_{|\mathbf{k}|<k_0} \left| \int \psi_{\mathbf{k}}^*(\mathbf{r}) f(\mathbf{r}) d^3\mathbf{r} \right|^2
\tag{A.5}
$$

Now let $k_0 \to \infty$, then the "best" value for $a_{\mathbf{k}}$ for $|\mathbf{k}| < k_0$ is when $a_{\mathbf{k}} = \int_{\Omega} \psi_{\mathbf{k}}^*(\mathbf{r}) f(\mathbf{r}) d^3\mathbf{r}$, which is unchanged as $k_0 \to \infty$.
Since $\Delta_{\min} \geq 0$ we get

$$
\sum_{\mathbf{k}} |a_{\mathbf{k}}|^2 \leq \int_{\Omega} |f(\mathbf{r})|^2 d^3\mathbf{r} \quad \text{Bessel Inequality.}
\tag{A.6}
$$

If $\sum_{\mathbf{k}} a_{\mathbf{k}} \psi_{\mathbf{k}}(\mathbf{r})$ is a continuous function of $\mathbf{r}$, then

$$
\Delta_{\min} = 0, \quad \text{if and only if} \quad f(\mathbf{r}) = \sum_{\mathbf{k}} a_{\mathbf{k}} \psi_{\mathbf{k}}(\mathbf{r}).
$$

Also

$$\Delta_{\text{min}} = 0, \quad \Longleftrightarrow \quad \sum_{\mathbf{k}} |a_{\mathbf{k}}|^2 = \int |f(\mathbf{r})|^2 d^3\mathbf{r}.$$

Thus

$$f(\mathbf{r}) = \sum_{\mathbf{k}} a_{\mathbf{k}} \psi_{\mathbf{k}}(\mathbf{r}), \quad \Longleftrightarrow \quad \sum_{\mathbf{k}} |a_{\mathbf{k}}|^2 = \int |f(\mathbf{r})|^2 d^3\mathbf{r}. \tag{A.7}$$

We say that $\psi_{\mathbf{k}}(\mathbf{r}, t)$ is a *complete orthonormal set* $\Longleftrightarrow$ for any continuous absolute square integrable function $f(\mathbf{r})$ for which $\sum_{\mathbf{k}} a_{\mathbf{k}} \psi_{\mathbf{k}}(\mathbf{r})$ is continuous.

$$f(\mathbf{r}) = \sum_{\mathbf{k}} a_{\mathbf{k}} \psi_{\mathbf{k}}(\mathbf{r}, t), \quad \text{where}$$

$$a_{\mathbf{k}}(t) = \int \psi_{\mathbf{k}}^*(\mathbf{r}, t) f(\mathbf{r}) d^3\mathbf{r}.$$

If the functions $\psi_{\mathbf{k}}(\mathbf{r}, t)$ of Eq. (A.1) are a complete set, then taking $f(\mathbf{r}) = \psi(\mathbf{r}, t)$, the wavefunction, we have

$$\psi(\mathbf{r}, t) = \sum_{\mathbf{k}} a_{\mathbf{k}} \psi_{\mathbf{k}}(\mathbf{r}, t), \quad \text{where} \tag{A.8}$$

$$a_{\mathbf{k}}(t) = \int \psi_{\mathbf{k}}^*(\mathbf{r}, t) \psi(\mathbf{r}) d^3\mathbf{r}. \tag{A.9}$$

Then for a normalized wavefunction $\psi(\mathbf{r}, t)$ we have

$$\sum_{\mathbf{k}} |a_{\mathbf{k}}|^2 = \int |\psi(\mathbf{r})|^2 d^3\mathbf{r} = 1, \tag{A.10}$$

which is the criterion for Eq. (A.1) to be a complete set.

A.2   Fourier Series

Consider a *single-valued, periodic function* $f(x)$ defined in the interval $-\pi \leq x \leq \pi$ and determined outside this interval by the condition $f(x + 2\pi) = f(x)$, so that $f(x)$ has the period of 2π. The *Fourier Series* corresponding to $f(x)$ is defined to be

$$f(x) = \frac{1}{2} A_0 + \sum_{n=1}^{\infty} \left[A_n \cos(nx) + B_n \sin(nx) \right]. \tag{A.11}$$

Provided $f(x)$ and $f'(x)$ are *piecewise continuous* in the interval $(-\pi, \pi)$, this series converges. The constants A_0, A_1, $\cdots$, A_n, $\cdots$ are determined by multiplying both sides of Eq. (A.11) by $\cos(mx)$ and integrating over the interval $-\pi$ to $+\pi$. We find

$$A_m = \frac{1}{\pi} \int_{-\pi}^{+\pi} f(x) \cos(mx) \ dx, \qquad m = 0, \ 1, \ 2, \ \cdots . \tag{A.12}$$

Similarly by multiplying Eq, (A.11) throughout by $\sin(mx)$ and integrating, we find

$$B_m = \frac{1}{\pi} \int_{-\pi}^{+\pi} f(x) \sin(mx) \ dx, \qquad m = 1, \ 2, \ \cdots . \tag{A.13}$$

Since $\cos(nx)$ and $\sin(nx)$ can be written as complex exponentials $\exp(\pm inx)$, the Fourier expansion Eq. (A.11) can also be written in the following form

$$f(x) = \frac{1}{\sqrt{2\pi}} \sum_{-\infty}^{+\infty} C_n \exp(inx). \tag{A.14}$$

The coefficients C_n can be found directly using the relation

$$\frac{1}{2\pi} \int_{-\pi}^{+\pi} \exp[i(n-m)x] \ dx = \delta_{n,m}, \tag{A.15}$$

where $\delta_{n,m}$ is the *Kronecker Delta Symbol* defined as

$$\delta_{n,m} = \begin{cases} 1 & \text{if} \quad m = n \\ 0 & \text{if} \quad m \neq n \end{cases} . \tag{A.16}$$

Thus multiplying both sides of Eq. (A.14) by $\frac{1}{\sqrt{2\pi}} \exp(-imx)$, integrating over x and using Eq. (A.15), we get

$$C_m = \frac{1}{\sqrt{2\pi}} \int_{-\pi}^{+\pi} f(x) \exp(-imx) \ dx, \tag{A.17}$$

where m is a positive or negative integer or zero.

A function $f(x)$ defined in some other interval $(-L, +L)$ and periodic with pertodicity $2L$, so that $f(x + 2L) = f(x)$ can also be expanded in a Fourier Series by making change of variable $x \rightarrow \pi x/L$. We then have

$$f(x) = \frac{1}{\sqrt{2\pi}} \sum_{n=-\infty}^{+\infty} C_n \exp[in\pi x/L], \qquad \text{where} \tag{A.18}$$

$$C_m = \frac{1}{\sqrt{2\pi}} \int_{-L}^{+L} f(x) \exp[-im\pi x/L] \ dx. \tag{A.19}$$

A.3 Fourier Transforms

Frequently, the functions with which we have to deal are not periodic, but are defined for all real values of x, $-\infty < x < +\infty$. Such functions can also be expressed in terms of complex exponentials by taking the limit $L \to \infty$ in Eq. (A.18) and Eq. (A.19). As L increases in value the difference between successive terms in the series Eq. (A.18) becomes smaller and smaller, and the sum over n can be replaced by an integral:

$$f(x) = \frac{1}{\sqrt{2\pi}} \int_{-\infty}^{+\infty} C_n \exp[in\pi x/L] \; dn. \tag{A.20}$$

Letting

$$k = n\pi/L \tag{A.21}$$

and defining a new function

$$g(k) = LC_n/\pi, \tag{A.22}$$

the integral Eq. (A.20) can be written as

$$f(x) = \frac{1}{\sqrt{2\pi}} \int_{-\infty}^{+\infty} g(k) \exp[ikx] \; dk. \tag{A.23}$$

By taking the limit $L \to \infty$ in Eq. (A.19) we find

$$g(k) = \frac{1}{\sqrt{2\pi}} \int_{-\infty}^{+\infty} f(x) \exp[-ikx] \; dx. \tag{A.24}$$

The integrals Eq. (A.23) and Eq. (A.24) are known as *Fourier Integrals*. And $g(k)$ and $f(x)$ are *Fourier Transforms* of each other. A function $f(x)$ can only be expressed as a Fourier Transform if the *infinite integrals* Eq. (A.23) and Eq. (A.24) converge. This will be the case if $f(x)$ and $g(k)$ are *square integrable functions*, which means that

$$\int_{-\infty}^{+\infty} |f(x)|^2 \; dx < \infty \quad \text{and} \quad \int_{-\infty}^{+\infty} |g(k)|^2 \; dk < \infty. \tag{A.25}$$

A.4   The Dirac Delta Function

By inserting $g(k)$ from Eq. (A.24) in Eq. (A.23) we get

$$f(x) = \frac{1}{2\pi} \int_{-\infty}^{+\infty} \left\{ \int_{-\infty}^{+\infty} f(x') \exp[-ikx'] \, dx' \right\} \exp[ikx] \, dk \qquad (A.26)$$

$$\text{or } f(x) = \int_{-\infty}^{+\infty} f(x') \delta(x - x'), \qquad (A.27)$$

where in Eq. (A.27) the order of the integrals has been reversed and the function $\delta(x - x')$ has been defined by

$$\delta(x - x') = \frac{1}{2\pi} \int_{-\infty}^{+\infty} \exp[ik(x - x')] \, dk. \qquad (A.28)$$

This procedure is open to question since integral in Eq. (A.28) does not converge. However, P.A.M, Dirac introduced this function attaching a meaning to it and it is known as *Dirac delta function*.

The main properties of the Dirac delta function are the following:

$$\text{(i).} \quad \int_a^b f(x) \delta(x - x_0) \, dx = \begin{cases} f(x_0), & \text{if } a < x_0 < b \\ 0 & \text{if } x_0 < a \text{ or } x_0 > b \end{cases}. \qquad (A.29)$$

$$\text{(ii).} \quad \delta(x) = \delta(-x). \qquad (A.30)$$

$$\text{(iii).} \quad x\delta(x) = 0. \qquad (A.31)$$

Actually the Delta function is defined by these properties. Any function that satisfies these properties is defined as Delta function. From these properties follow others.

$$\delta(ax) = \frac{1}{|a|} \delta(x), \quad a \neq 0. \qquad (A.32)$$

$$f(x) \delta(x - a) = f(a) \delta(x - a). \qquad (A.33)$$

$$\int \delta(a - x) \delta(x - b) \, dx = \delta(a - b). \qquad (A.34)$$

$$\delta[g(x)] = \sum_i \frac{1}{|g'(x_i)|} \delta(x - x_i) \qquad (A.35)$$

$$\text{where } g(x_i) = 0, \quad \text{but } g'(x_i) \neq 0.$$

$$\lim_{a \to +\infty} af(ax) = \delta(x), \text{ if } \int_{-\infty}^{+\infty} f(x) \, dx = 1. \qquad (A.36)$$

Example A.1

$$\delta\left[\left(x - a\right)\left(x - b\right)\right] = \frac{1}{|a - b|}\left[\delta\left(x - a\right) + \delta\left(x - b\right)\right], \tag{A.37}$$
$$\text{for} \quad a \neq b$$

Example A.2

$$\delta\left(x^2 - a^2\right) = \frac{1}{2|a|}\left[\delta\left(x - a\right) + \delta\left(x + a\right)\right]. \tag{A.38}$$

Example A.3 Since $\frac{1}{\sqrt{\pi}}\int_{-\infty}^{+\infty}\exp\left[-x^2\right] = 1$, so

$$\lim_{a \to +\infty} \frac{a}{\sqrt{\pi}}\exp\left[-a^2 x^2\right] = \delta\left(x\right). \tag{A.39}$$

The derivative of the delta function

$$\delta'\left(x\right) = \frac{d}{dx}\delta\left(x\right), \tag{A.40}$$

can also be given a meaning, since

$$\int_a^b f\left(x\right)\delta'\left(x\right)dx = \left[f\left(x\right)\delta\left(x\right)\right]_a^b - \int_a^b \delta\left(x\right)f'\left(x\right)dx$$
$$= -f'\left(0\right), \tag{A.41}$$

where the $x = 0$ point is assumed to lie in the interval (a, b). Otherwise

$$\int_a^b f\left(x\right)\delta'\left(x - x_0\right)\,dx = -f'\left(x_0\right), \quad a < x_0 < b. \tag{A.42}$$

We have further properties of Fourier transforms from Eq. (A.23)

$$\int_{-\infty}^{+\infty}|f\left(x\right)|^2 dx = \frac{1}{2\pi}\int_{-\infty}^{+\infty}dx\left[\int_{-\infty}^{+\infty}g^*\left(k\right)\exp\left(-ikx\right)dk\right] \times$$
$$\times \left[\int_{-\infty}^{+\infty}g\left(k'\right)\exp\left(ik'x\right)dk'\right]$$
$$= \int_{-\infty}^{+\infty}g^*\left(k\right)dk\int_{-\infty}^{+\infty}g\left(k'\right)dk'\left[\frac{1}{2\pi}\int_{-\infty}^{+\infty}\exp\left(i\left(k - k'\right)x\right)dx\right]$$
$$= \int_{-\infty}^{+\infty}g^*\left(k\right)dk\int_{-\infty}^{+\infty}g\left(k'\right)\delta\left(k - k'\right)dk'$$
$$= \int_{-\infty}^{+\infty}|g\left(k\right)|^2 dk. \tag{A.43}$$

This result is known as *Parseval's theorem.*

One can generalize Fourier's transforms to functions in three dimensions:

$$f(\mathbf{r}) = \frac{1}{(2\pi)^{3/2}} \int g(\mathbf{k}) \exp[i\mathbf{k} \cdot \mathbf{r}] \, d^3\mathbf{k}$$

$$g(\mathbf{k}) = \frac{1}{(2\pi)^{3/2}} \int f(\mathbf{r}) \exp[-i\mathbf{k} \cdot \mathbf{r}] \, d^3\mathbf{r} \qquad (\text{A.44})$$

Bibliography

[1] B.H.Bransden & C.J.Joachain: *Introduction to Quantum Mechanics*, (Longmans, ELBS Edition, Harlow, Essex, 1990)

[2] A.S.Davydov: *Quantum Mechanics*, (Pergamon, London, 1965)

[3] P.A.M.Dirac: *Quantum Mechanics*, 4th Edition, (Oxford Press, Oxford, 1958)

[4] P.K.Ghosh: *Quantum Mechanics*, (Narosa, New Delhi, 2014)

[5] K.Gottfried: *Quantum Mechanics*, (Benjamin/Cummings, Reading, MA, 1966)

[6] E.Merzbacher: *Quantum Mechanics*, 2nd Edition, (John Wiley, New York, 1970)

[7] Nandita Rudra & P.Rudra: *Basic Statistical Physics*, (World Scientific, Singapore, 2010)

[8] Lewis H.Ryder *Quantum Field Theory*, (Academic Publishers, Calcutta, 1989)

[9] J.J.Sakurai: *Modern Quantum Mechanics*, Revised Edition. (Addison-Wesley, Reading, MA, 1999)

[10] L.I.Schiff: *Quantum Mechanics*, Third Edition,(McGraw- Hill, New York, 1968)

Index

Printed and bound by CPI Group (UK) Ltd, Croydon, CR0 4YY

17/10/2024

01775694-0008